squills

THE INTERNATIONAL PIGEON RACING YEAR BOOK

1993

'OLD 86'

EDITED BY ''SQUILLS''

95th YEAR OF ISSUE

The list of contents and the index of
advertisers appears at the back of this edition.

ISSN 0952-4541
ISBN 0 85390 037 X

Squills International Pigeon Racing Year
Book, incorporating Squills Diary,
Squills Annual and Squills Stud Book,
and the Pictorial Pocket Diary.

© The Racing Pigeon Publishing
Co Ltd 1992

Origination by RP Typesetters
Phone: 071-833 3201.

Printed by Unwin Bros,
Woking, Surrey.

Published by The Racing Pigeon
Publishing Co Ltd, Unit 13,
21 Wren St, London WC1X 0HF.
Phone: 071-833 5959 Fax: 071-833 3151.

INTRODUCTION

By the time this introduction is written I have had the privilege of reading the articles that make up this year book and as always they are a mixed bag because the best fanciers are not always the best writers, indeed one of our biggest winners says of himself that he can neither read or write. Nevertheless because he is a very good fancier it furnishes an outstanding article. Other articles are short but even the shortest has the gold nuggets of information in them. For novices sometimes these are the best articles because the winning fancier gets to the point quickly. The novice is in search of that magical ingredient that will make him successful. The difference between success and failure is often something so small that it might otherwise be missed.

The fancier who is always down the list will want to know what he is doing wrong. By reading any of these articles he could so easily come across that little item he had missed or even that he had forgotton. This revelation might come from one of the shorter articles or one of the longer ones. Some of these long stories have their own attraction because they can go into a lot more detail and this in itself is interesting because the novice will be able to follow a successful system right the way through and this will be of great use when he (or she) comes to work out a system of winning for themselves.

This is the secret of getting the most out of Squills. Read all the articles, since they are all written by top winners with the novice in mind, they are all worth reading. As I wrote last year, this is one of the wonderful things about pigeon racing that so many successful fanciers are so willing to help novices.

I was also pleased at the number of fanciers who wrote or phoned to say that I had hit the nail right on the head with my comments about buying pigeons even though I was only repeating what I had been saying for half a century and more. This year, in case anyone has not noticed, is the year when European unity has been on the lips and in the minds of the great majority. For pigeon fanciers we can look forward to some things getting easier but what and when is another matter.

On quarantine at the time of writing there is no change. A phone call to the Ministry of Agriculture on 6 October 1992 found out that there were no changes planned so that the article DIY Quarantine published first in the RP Pictorial is still valid. As we said at the time if three or four fanciers are visiting the Continent if one of them clears his loft putting the birds somewhere else he can turn his own loft into a quarantine station for himself and his friends. It needs a bit of planning but if you want to import birds it will save a lot of money.

I have an open mind on whether all these imports are that much better than true-blue British birds but on the other hand I am all for fanciers having a few days holiday, visiting a few Continental fanciers and bringing back a few birds for the fun of it. It may not be scientific breeding but could be the start of many international friendships.

The dawn of the European Community has actually reached Squills as can be seen by the number of advertisements from Belgian fanciers. Some of these arise out of what has become an annual event, our Belgian Safari, where RP staff under the command of Tony Oates take the boat to Ostend and based in the hotel owned by RP and Pictorial correspondent Eric Deledicque collect birds for the Old Comrades (sorry British National Show).

The Belgian birds so generously given are one of the important sources of money for the disabled ex-servicemen's home. The British contribution is equally important but because it does not involve quarantine is that much easier. This is another heart-warming aspect of the sport, the great generosity of fanciers for charity and for novices a fine antidote to the greed and commercialism unfortunately shown by a few.

AN AMBITION REALISED

by George Campbell from Annan
1st SNFC Sartilly (2)

It was my good fortune during the season past, to time from SNFC Sartilly (2) to be 1st South Sect, 1st Open and thus realise a life-time ambition — to win the prestigious Scottish National. Following my National win I was asked to compile an article for this famous Squills Year Book and trust that our readers find this contribution of interest. As long as I can remember I have had an interest in pigeons and when only a schoolboy enthusiastically visited local fanciers to help them with their birds. I used to visit the loft of Percy Moore to help him to clean his loft and care for his pigeons. Percy was the brother-in-law of the late, great John Kirkpatrick and actually lived in the former home of this legendary fancier. Whilst I did not appreciate the fact at the time, Percy had pigeons closely related to all the Kirkpatrick champions, which enabled me, in these schoolboy years to work with quality bloodlines.

My parents lived local to the homes of several prominent Annan fanciers. I visited the home of the late George Jackson who was an accomplished administrator, being the former secretary of the Solway Fed, but he was also a very talented fancier owning a number of fine pigeons and had been 2nd Open SNFC Nantes. I was a regular visitor to the home of Bert Tennant who was to encourage me with stock and advice. Bert currently competes in partnership with Eddie Anderson and together they have won two Nationals. I shared a friendship also with the late Peter Grieve and recollect assisting him when he moved the loft from a piece of common ground in the town into his back garden and settled the pigeons. It was this practical involvement which kindled my interest in the pigeon Fancy. When still at school my parents arranged for a small loft to be built in our back garden and I joined the famous Annan and District HC. My first winner was a light chequer which I had received as a gift from Thomson Bros. Following my marriage I moved to live in a different part of the town and a loft was erected in the garden of our new home. At that time the loft title was changed to include my father's name. Building upon the original lines we assembled a most useful team and scored consistently from the Classics.

In 1975 I timed a blue chequer cock to be 1st Sect, 4th Open SNFC Rennes and incidentally the previous year this same pigeon was 25th Sect, 64th Open SNFC Avranches, when two years old. The dam was home bred from a hen which I received from the late Dave Shaw when paired to a cock I received from fellow club member Billy Dalrymple through an introduction which Billy made from Ayton Marshall from Seaton Sluice. The sire of the chequer cock which won the section from Rennes was received as a gift from Irving & Johnstone who were, of course, the breeders and racers of the immortal 'Solway King'. Jimmy Irving and

George Campbell

Eric Johnstone were to subsequently split partnership, but coincidentally, Eric had an influence in a further good pigeon which I owned. In 1976 I was 2nd South Sect, 2nd Open SNFC Nantes being beaten by Walter Halliday from Brydekirk. My winning pigeon was a red chequer hen, the dam of which was home bred from a hen which I received from George Jackson when paired to a red black splashed cock I obtained from Eric Johnstone which had been bred by the late Willie Graham from Annan. Willie assisted John Kirkpatrick with the management of his pigeons during his final illness and assisted Mrs Kirkpatrick to care for the team following her husband's death. Willie Graham had the absolute cream of Kirkpatrick bloodlines and the red cock was through these lines. The sire of the red hen which was 2nd Open Nantes was a mealy cock obtained from friend Bert Tennant, containing the Kirkpatrick bloodlines of Thomson Bros. In 1976 also, I won the magnificent Ellsworth Trophy for the best average from Rennes and Nantes with nominated pigeons and The News of the World Trophy for the best average from these two prestige events and I vividly recollect how proud I was when I stepped forward at the Scottish National prize presentation to accept these trophies.

We then parted with the team and I was out of our sport for several seasons although always retaining an interest. In 1983 I decided to re-enter the Fancy and during the summer months erected a new loft in the back garden. I share a close friendship with my near neighbour, James Dalgliesh, and as the young bird season drew to a close in 1983, James & Dylis Dalgliesh went on holiday and I agreed to care for their pigeons and race them from the final two races. It was a great pleasure for me to be practically involved attending to a team of pigeons once more and particularly to care for such a quality team. When James returned home he told me to get a basket out of the shed and to my surprise that day gifted to me seven performance pigeons to start me away again. What an opportunity this was and over the subsequent years I have worked away with these base bloodlines, supplemented with introductions from friends, pigeons from Eric Johnstone and Dickie Graham having a particular influence. Among the pigeons obtained from James Dalgliesh were two mealy cocks, one being the sire of the other and the older of these two pigeons had scored three times from the Channel for the breeder. During my first season back in the Fancy, the younger of the two cocks was my only entry from the Channel from Sartilly and was pleasingly up with the leaders to be 31st Sect, 102nd Open. This whetted my appetite for further competition!

The loft which I had built is 15ft long, with a flat roof containing roof lights which enables the sun to shine on to the floor in front of the boxes. The nest boxes are conventional Natural boxes closed-in at the front.

The smaller of George Campbell's lofts which has been adapted for Widowhood.

Ventilation is through a gap at the top front and through louvred windows
in the end of the shed. This loft has since been extended in width and
is now 8ft wide. The pigeons were originally raced on the Natural system
with reasonable success. I noted, however, the success being achieved
in England on the Widowhood system and whilst desiring to retain the
hens on the road, felt I should look towards modifying my system. I
therefore built a further 7ft x 5ft loft into which I could race the youngsters
and modified the original loft to enable me to race part of the team on
the Roundabout system. I have persevered with the original bloodlines
and several useful pigeons have emerged. A home bred 1986 blue cock
raced particularly well. He was 10th Sect, 53rd Open SNFC Sartilly (1)
1989; 15th Sect, 25th Open Rennes 1990; was intended for Rennes again
in 1991 but I was not satisfied with him on basketing for Rennes and held
him for Niort when he was 11th Sect, 12th Open. He was raced Natural,
paired to a barren hen and was dummied to set him up for the Classics.
The sire was one of the original mealy cocks from James Dalgliesh and
the dam was a blue hen which friend Ewart Warwick purchased for me
from a fancier in Dumfries who was leaving the Fancy.
   A 1987 home bred chequer cock was 16th Sect, 42nd Open SNFC
Sartilly (1) 1990 and the same season was doubled back into Niort to be
12th Sect, 45th Open. He was raced Roundabout and paired for the
Channel. The dam is a blue chequer hen sired by a pied cock which won
Cheltenham and goes back to James Dalgliesh's pied hen that was 2nd
Sect Rennes. The sire is a mealy cock, the grand-dam of which is the
dam of 'Annandale Rose', James Dalgliesh's SNFC Gold Award winner.
I was fortunate to receive as a gift a 1986 blue chequer hen from Dickie
Graham from Ecclefechan which contains Stassart x Kirkpatrick
bloodlines. She went to the Channel twice in 1988 to be 29th Sect Sartilly
(1) and 11th Sect, 56th Open Sartilly (2). In 1989 she raced from Rennes
and was 36th Sect, 94th Open and returned to Rennes in 1991 to be 35th
Sect, 88th Open. She races on the Natural system to a small youngster
but as she has bred several promising pigeons and a son has scored twice
from the Channel it is my intention to use her extensively at stock.
   Being motivated by the talented National-minded fanciers around me
my objective is to succeed from the Channel and with this in mind do
not pair my Natural pigeons until late March. Due to restriction of time
I was not able to put the team on Roundabout during the season past but
when on the system they would be paired slightly earlier. The birds are
encouraged to build their own nests using dried reeds which I gather on
the Solway shore. When making my pairings I put best to best racers
with a consideration of type. Throughout the season I consistently feed
a no-bean mixture with Bamfords 'Eurowinner' added. I can be considered
a heavy feeder as the birds are hopper fed with corn in front of them
at all times. Red Band and hemp are used to induce condition as the birds
are being prepared for the Classics. I house a modest team, wintering
22 pairs, including two pairs of stock pigeons. I breed 24 youngsters which

The interior of George Campbell's primary loft.

are raced with caution having only three or four young bird races. As yearlings I put them to 25 miles and then as two year olds they must go to the Channel. As preparation for the Classics they have three or four inland races and are then set-up for the chosen event.

I am always observing and try to enter the Channel candidates in their keenest nest condition but generally speaking prefer cocks sitting seven-to ten-day eggs and hens racing to a four- or five-day-old youngster. The team is exercised around the loft morning and evening for 45 minutes but I emphasise that any birds on the nest are not lifted off to exercise them as it would break their concentration. In addition to loft exercising them they are allowed the open hole all day. I am fortunate that I now possess a car which allows me to work the birds on the road for the Channel and as the Classics approach they are tossed as often as possible from 25 miles. The smaller of my two lofts has recently been modified to contain Widowhood nest boxes and I am considering trying several cocks on Widowhood some time in the future.

My ambition has always been to win the Scottish National and I was fortunate to realise this ambition from SNFC Sartilly (2) in 1992. Following a two-day holdover the convoy was liberated at 8.30am on Sunday July 19 into a fresh SW wind and I timed at 5.31 to record 1422 ypm. When the clocks were opened at the local clock station on the night indications were that I had won the South Sect but I had to wait for it to be confirmed as the winner of the Open event. It is difficult to describe my elation on realising a life-time ambition. I must not forget the pigeons which were just behind me in the result, particularly the good pigeon belonging to Jeff Horn which was 2nd Open. My winner is a three year old blue hen unraced as a youngster and raced to Mangotsfield as a yearling. She went to Rennes as a two year old and was 25th Sect, 70th Open, racing to a five-day-old youngster. She was my fancy again for Rennes this season and was being prepared with the Blue Riband event in mind.

I would normally race my Rennes candidates to Cheltenham only before entering them from the Channel but we experienced very easy inland racing to Cheltenham and I believed another race from Mangotsfield would just put the edge on the pigeon. Mangotsfield proved a very difficult race with a large percentage of the convoy lying out and unfortunately the blue hen did not return until the next day. I see no sense in entering pigeons for Channel events unless they are right and just as I held the blue cock for Niort in 1991 I decided immediately not to send the blue hen to Rennes but to hold her for Sartilly (2) when she would be right. She was basketed on a five-day youngster with her second flight full up in one wing and her third flight half-grown in the other. The dam was bred by Jim Muir from a hen introduced from Top Lofts from their 'old family' when paired to a Vanhee from Mr & Mrs Clive Bryson from Gretna. Jim, incidentally, is married to my niece Susan. The sire of the blue hen is a blue cock which came to the loft as a flyaway and when reported belonged to Rae & Gardner from Brydekirk. John Gardner worked beside me and allowed me to keep the pigeon and he proved "a real one" scoring in the club and with the SNFC from Sartilly. The blue cock was bred from a home-bred hen when paired to a Janssen cock which Alistair Rae & John Gardner obtained from Billy Swan from Dumfries.

The National winner is a real favourite in the loft playing up to me when I speak to her and being very affectionate. As I explained the birds build their own nests and when the blue hen goes to nest she carries nest materials for days from gardens in the streets where we live and from the large playing field which is sited behind our home. I am employed as a painter which requires me to travel from home, particularly during the winter months and I am fortunate that my good lady, Ann, cares for the pigeons while I am away from home. Without her assistance I would find it difficult to manage the team and in appreciation have named the National winner — 'Lady Ann'.

I thank fanciers and friends who offered congratulations, following our National triumph and thank the friends who have assisted me over the years with stock and advice. To them and to all fanciers I wish the best of luck for the year ahead.

RACING PAIRED HENS

by J Pilling, Skelmersdale
The star writer of 'Pillings Patch'

A little while ago a friend of mine casually mentioned that it might be a good idea to race hens which are paired together so I sat down and worked out how I would use such a system and although I am no longer able to race I see no reason why it should not be successful. Just as for any method a fancier needs good pigeons and on top of that they must be healthy and fit for without these requirements the best method in the world would not succeed. Although I am writing with the small team fancier in mind this method will do for any number of pigeons. Unfortunately there is one drawback and that is time; or to be more precise, the lack of time and it is difficult to see how a working fancier can put the idea to use because these hens would have to be exercised and trained separately from any other pigeons in the loft so a fair amount of time is needed.

Many fanciers are reluctant just to race cocks on Widowhood because it is a waste of good hens so they race on the Natural system but in doing so it very often happens that they fail in the early races and it is the Widowhood cocks that shine. There are exceptions of course and pigeons on Natural have won in the early races but more often than not it is when the weather is not too good but this is only my opinion. On the other hand fanciers with only a small team of Widowhood cocks are struggling to have something in form after the first six or seven races and it is then that the Natural pigeons show their worth. Generally speaking whether racing a small team of Widowhood cocks or racing on Natural there is some stage of the programme when the birds are not quite right as far as winning is concerned. Hopefully this idea will enable a fancier to have something in form for every race — a good hen is hard to beat in the shorter races when the weather is not too good and hard to beat in the longer races no matter what the weather conditions are. I will stick my neck out and say that hens raced on this system will be hard to beat in the shorter races even if the weather is good.

The system requires a fancier to have a small team of Widowhood cocks to be raced on the Widowhood system. It also requires a small team of hens and for my explanation we will assume that a fancier has 12 such hens. It is the management of these hens I will explain, bearing in mind that the Widowhood cocks should be on song for the first six or seven races. In these races the cocks will be backed up with the hens so in these races you have a fairly strong team and the hens will also compete in the longer races as well. Only a small percentage of pigeons are capable of flying long hours at a racing speed even if they are healthy and fit. So these hens must be the right type, and for me if a pigeon has what

I call the "Bilco" wing and is healthy and fit and in a good nesting condition it has every chance in the longer races.

Using 12 such hens, the loft only needs to be 8ft x 6ft with a partition in the middle and a door but no gaps in the partition so that pigeons cannot see into one section from the other. In one section would be six nest boxes and these only need to be the type which are used for pigeons on Natural. There would be three rows of two with the bottom row raised off the loft floor. A 6in wide board would run along each row enabling a pigeon to fly on to this before entering the box and it also serves as a perch when the boxes are closed. There would also be a 6in wide board which is upright between each box, so in effect the front of each box looks like a large box perch. The front of this section would have vents at the bottom and also vents at the top rear, also it would have a trap and a landing board 4ft long. The remainder of the front would have some perspex to let in light so that the pigeons cannot see outside. The front of the other section would have a door to gain access to the loft but no landing board or trap. It would also have some perspex and vents front and rear. The inside of this section would have about eight V-perches and the floor covered with a strong mesh grill, plus a raised-up drinker in each section.

About the end of January the 12 hens are put into the loft with the partition door open and all nest boxes open. It may seem too early to do this seeing they will be racing in the later races but it will take them quite a while to pair together so I think the end of January is just about right. They are confined to the loft for a few days then for another two days are fed sparingly and then let out through the door in the perch section. The trap in the other section is opened and the door closed. Being hungry they will fly for a short time and with a bit of trapping seed they can be tempted on to the landing board and enter the trap. After a few days of doing this they learn the drill of coming out through the door and entering by way of the trap. If possible exercise once a day but this may be difficult as the days are short at that time of the year.

The idea is to get these hens to pair together letting them choose their own partners and their own nest boxes. This can be encouraged by feeding the same two hens in the same box once you have seen one or two of them entering the boxes. They will not all pair up at the same time but with a bit of encouragement they will pair and now is the time to put nest bowls and straw in the boxes. Eventually some will lay and in some nests there could be four eggs, while in others two, but I think if only one of a pair lays the other will sit the eggs its partner has laid. There is no rush to have them sitting but it is better not to leave it too long if some are slow to lay they can be encouraged to sit by placing a warm pot egg in the bowl in the evening and if they take to it put another one in two days later. If any eggs look to be thin shelled replace them with pot eggs because you do not want eggs to get broken and the pigeons leaving the nest. It might be better to replace all eggs with pot eggs. If you do leave eggs in the nest make sure they are not fertile just in case a hen has been flirting about with a cock outside because on no account must they see a young one.

They are allowed to sit as long as they want and with eggs not being fertile some will sit up to a month. Once a pair have left the eggs one of the pair is put into the other section and the nest box closed. This is important. As each pair give up sitting one of the pair is put into the other section with the nest boxes closed so you end up with six hens in each section. During the sitting period give them a few tosses up to about ten miles. It would be better if only one of each pair is trained at a time, ensuring they all get trained. The reason for this is if all were trained at the same time the eggs would get chilled which could result in some of them refusing to go back on them. If it seems I am spelling all this out rather slowly it is because I think even the small things in pigeon racing are important.

We now have the situation where all hens have sat out their time and both groups of six in the two sections. It is important from now until part-way through the racing season that each group occupy the perch section for two days at a time then change them over and continue doing this. One reason for doing this is if one group was left in the box section all the time they may pair together and this would upset the whole idea. Another reason is that it is rather uncomfortable in the perch section and they will be keen to get back into the box section and this is one more way to motivate them when racing starts. For the first few days of separation they are not let out and after that they are exercised and trained.

To exercise, let those in the perch section out through the door then shut the door. Open the trap on the other section and drive the section group into the perch section. When the first group have exercised they will trap into the box section and then the other group are let out. While the second group are exercising drive the first group into the perch section and when the second group have exercised they trap into the box section and you end up with the same six pigeons in the same two sections; just like the Roundabout. The idea for separate exercise for the two groups is to prevent them seeing each other too often and may I remind you to let each group occupy the perch section for two days at a time.

During separation they are trained and it should now be about three weeks before the first race so it is important to plan the date when they are first paired up and as mentioned I think the end of January is about right. To train them during separation put one group in one basket and the other group in another basket and let one group go at least a quarter-of-an-hour before the other then when you get back separate the two groups again, so for these training tosses they do see each other but only for a short time. They would need about ten tosses like this from five miles to 40 miles and then should be ready for the first race and may I add that the Widowhood cocks should also be ready.

On the Friday before the first race they are fed in their separate sections about noon, then half-an-hour before taking them to the marking station the door in the partition is opened and all nest boxes opened. After being together for half-an-hour you then basket the group which came out of the perch section and these are sent to the race.

The other group are not raced that week and are in the box section when the racers return. After the race they are again left together for half-an-hour then separated. The following week the same thing is done but this time the other group are raced, making sure that this group are in the perch section on the Friday when the partition door is opened. So you see that each group is only raced every other week so they are not over raced when the distance races are flown. In between each race one group is given a 40-mile toss on Wedneday and the other group the same on Thursday. For these tosses they do not see each other, and both groups are exercised separately each day of the week excepting of course on Friday and Saturday.

It is up to each fancier to decide how many races are flown like this bearing in mind that they will also compete in the longer races. Fanciers can make their own plans for the distance races. As an example you can re-pair two of the pairs long before the first distance race to have them sitting about ten days for the race and all four are raced. When they home from the race coax them back on the eggs and if you warm the eggs they will do it. For a later race put a six-day-old young one in the nest 36 hours before basketing. For the second longer race re-pair two more of the pairs to be sitting ten days and for a further race use a young one. For the third longer race re-pair the last two pairs in the same way and by using eggs and young ones you can get two good distance races out of each of the 12 hens, not forgetting they will be in the right frame of mind in the shorter races as well, but this sitting or seeing a young one is held back for the longer races. The advantage of using this method is that all 12 are raced and all are motivated to gain that extra bit of speed. When they are re-paired and sitting for the longer races they are keen as they have been deprived of eggs for a long spell and extra keen on a young one as this will be the first young one they have seen all year.

The object of this article is to explain the racing of hens so I will not go into great detail about the feeding or how to keep them healthy but in pigeon racing these two things are just as important as the method of racing; perhaps more so. For my part I would feed them as for Widowhood and I would bear it in mind that there could be a tendency to overfeed, seeing they are only exercised once a day and only raced every other week. If more exercise is given or more training they will have to be fed accordingly. With only racing every other week I think I would feed a good quality barley for four or five days after a race, but no matter how much food is given, it should be split into two feeds, morning and evening. Pigeons which are only fed once a day in the evening are hungry when let out for exercise in the afternoon and hungry pigeons will not exercise well.

There is no doubt that the most important single factor in pigeon racing is the health of the birds and I believe that many are sent to races with no chance of winning because they are not quite right in themselves. Many good fanciers say they do not treat for anything unless there is something visibly wrong and I respect their views but I do not agree with them on this matter. I have to admit that I cannot tell if a pigeon is healthy by just looking at it unless of course something is very much wrong and I suspect that thousands of fanciers are in the same boat. A pigeon can have all the looks of being healthy and yet it could have a mild form of some kind of ailment and it takes two or three weeks of poor performance before some fanciers realise that something is wrong. Then there is the big problem of knowing what the ailment is and by the time the pigeons are got right, racing is almost over for that year. Many years ago I only treated against worms and cocci and one year they were not racing as well as I expected so I took four pigeons to a very good vet and was told that all four had too high a canker count and from that day I had a set programme for health treatments for the four common ailments. This does not mean that antibiotics are pumped into them week after week; commonsense has to prevail as we all know that with too much antibiotics the pigeons become less resistant to diseases.

I did not intend to dwell on the health issue so back to the hens and if I used this idea I would use yearlings for various reasons. First they would be much easier to settle to the loft than older hens. Secondly it has been proved that the right type of yearling is quite capable of winning up to 500 miles. Perhaps the third reason is the most important in that yearling hens are very keen when sitting eggs. It is something new to them and extra keen on a young one as this would be the first young one they have had in their care. Perhaps some brave fanciers will have the courage to try it and I am confident they will not be disappointed providing the birds are the right type for the distance and are healthy and fit.

SUCCESS FROM THE START

by Lilley & Son of Hartlepool
Winner of 1st Up North Combine Provins (567 miles)

We would like to thank the Editor for giving us such an opportunity as this to write an article for 1993 Squills Year Book. Racing pigeons go back three generations in the Lilley family, my father first kept pigeons with my grandfather David Lilley in 1946 under the name Lilley & son, flying in the Hartlepool HS. After he died in 1960 my father kept them for a further year, but with a young family, he decided to give them up. My younger brother, sister and I had always visited the allotments, so it was only natural that one of us would become interested.

My first pigeons were kept in the coalhouse out in the back garden and I took great pride in these even though they were fancy pigeons. As the hobby increased, my father built me a small loft at the bottom of the garden. After leaving school and getting my first job as an apprentice fitter, he decided to get back into pigeon racing. This was very different for me from fancy pigeons, but my father was a good teacher. Our first loft was small, but it was the start of Lilley & son. 1974 flying in the Hartlepool Workingmen's HS we started with gift birds of Vandies, Kirkpatrick strain from my uncles, Lilley Bros. Also purchased were four Barker birds off Bill Orley, former UNC winner of Hartlepool.

Our first season in 1975 was racing broken old birds. We bred a good young bird team, this showed in the results. The first Selby YB was won with a grizzle Kirkpatrick cock. This cock also won the next Selby YB race, so we had the taste of success from the beginning. A 3rd Grantham and 2nd Ashford National, finishing the season runner-up YB Av and this in our first year racing. 1976 we raced OBs, mainly yearlings, these birds flew very well gaining some good club positions. One bird, a blue Vandy cock was 2nd Club Bourges, 553 miles, 108th UNC. That year we won Channel Av, Combine Av and YB Av.

Our team was increasing quickly so we had to think of a bigger loft. My uncle's allotment had space for us to build a bigger loft. So we raced OBs to the smaller loft, while building at the new location ready for the YBs. Young birds flew very well to this new location. At the end of the season we moved all our birds into the new loft. This was built to our own design; two compartments, 15 boxes for Natural racing, 30 box perches for youngsters. 1978 brought some excellent performances, racing to this new loft: 1st Club, 1st Fed Lillers, 47th UNC with a chequer Vandy hen; 1st Club, 2nd Fed Bourges, 25th UNC, 553 miles with a blue Barker hen bred by Bill Orley; 1st Club Ashford, 4th UNC Wednesday Club National, with a red Kirkpatrick cock. We won Channel Av, Combine Av runner-up; Fed Channel Av; this made us top prize-winner in the club.

During 1979 season we purchased an allotment at the rear of my father's

'Champion Sealink'

house. At the end of the season saw us building another loft, 24ft in length, 6ft in depth, this is our present location. New stock was brought in. These were Busschaerts. 1980, all birds had settled to the new loft quite easily. We decided to fly some birds on the Widowhood system and this proved to be successful. So in 1981 we put 12 cocks on Widowhood and 12 pairs Natural. The Widowhood cocks were a mixture of Busschaert, Kirkpatrick and Vandy. The Busschaerts flew better on this system. Over the next few years we gradually increased the strain of Busschaert, while phasing out the Vandy and Kirkpatrick, gaining many club and Fed positions and Open races.

In the middle 1980s some new breeds were having lots of success. These were F V Wildemeersch and Janssen so we decided to introduce these to our loft. Reading many times about a fancier from the Midlands — Len Hopton — I was able to purchase two pairs of Janssens from him. Next we saw that Bert Fletcher was having a sale at Haswell Plough. Here we bought a pair of young F V Wildemeersch which later turned out to be two hens. Two cocks were purchased to go with these hens. Both the F V Wildemeersch and Janssens breed many winners. 1988 we introduced into the old Busschaert blood some new Busschaerts from Hindhaugh & Donaldson, Tom Larkins and Bert Fletcher. The present team of birds consist of these three families which have bred many winners of club, Fed and Combine positions.

The Combine winner champion 'Sealink' is a F V Wildemeersch and has always been raced on Widowhood. His dam is one of the original hens purchased in 1985 from B Fletcher and his sire was from F Faulkner. A late bred in 1988, he raced steadily the following year. As a two year old he went to the Channel, first race Abbeville, 326 miles, 10th Club. Two weeks later Clermont, 400 miles, 1st Club, 1st Fed, 1st Sect, 15th UNC, 10,533 birds. Rested four weeks, sent to Provins, 467 miles, where he was 2nd Club, 2nd Fed, 3rd Sect, 2nd FCC, 2nd NECC, 17th UNC, 6,520 birds. Another two weeks later going to Abbeville, 326 miles, 2nd Club, 16th Fed, 227th UNC, 14,828 birds. For the above performances he won best bird NEHU 1990 plus Sealink Trophy, hence the name 'Sealink Cock'.

The following year 1991 he scored from Provins, again being 2nd Club, 14th Fed, 120th UNC. In 1992 we gave him two short races then put him into stock loft for a round of eggs. Returning him to the race team, first race Folkestone, 268 miles, then two weeks later we sent him to Abbeville where he was 9th Club, another two weeks going to Clermont. Next Channel race being four weeks to Provins. Looking in tiptop condition as always, with each race, we thought this will be the race we achieve our ultimate goal. Pigeons went on the Thursday. Birds were held over on Saturday. Sunday, weather hopefully seemed better, pigeons were liberated 6.40, north wind, hard race expected.

Standing at the loft with anticipation, the 'Sealink Cock' dropped in after 11 hours 57 mins on wing, looking in excellent condition and 25 minutes in front of next bird in club. On arriving at Fed marking centre, we found that we had the best bird in Hartlepool Fed. Later that night the Fed secretary phoned to inform us that it was looking hopeful in the Combine. Phoning different people that we knew in Combine, to see if they knew of any better birds we heard there was a good bird in at Tynemouth. I thought to myself "nobody remembers a 2nd". Next day at 11 o'clock we got word that we had topped the UNC. The feeling was of great satisfaction after all the years of racing and the hard work, we had achieved it.

On closing this article we would like to thank the following fanciers for helping us build a team of consistent racers inland and Channel: Len & Barbara Hopton, Bert Fletcher, Frank Faulkner, Hindhaugh & Donaldson, Tom Larkins. Also we would like to thank members of Hartlepool Fed, Hartlepool Workingmen's HS on congratulating us on winning the UNC.

THE BARCELONA DOUBLE

by Geoff Hunt of G Hunt & Son
Twice winner of BICC Barcelona

Although keeping pigeons since 1938 it was not until the autumn of 1945 that I began to meet some of the top fanciers of the south east area of Kent. Some of these were miners that were able to fly inland races (300 miles) during World War Two and others that had been in the Services and were now back in 'civvy street'. Some of them had been serving with the National Pigeon Service, and now had their own teams of birds to race again.

It was from these men that I listened and began learning all about distance racing. Race points always being mentioned were the likes of Marseille, San Sebastian and Barcelona. Within a very short space of time, I only lived for racing pigeons and my dream and challenge in the sport was to try and time alongside some of the great names that had become household ones in the National and extreme distance races.

In the beginning I received gift birds and also purchased a few Logans for 10/- (50p) each, one of which won my first young bird race entered in, winning 1st Club, 1st East Kent Amal at 1428 yards per minute in 1946. I also had a few youngsters from John Banks of Folkestone, Funny Colour strain, and one from Jim Bruton of Palmers Green, a brother to Bruton Supreme. So the long road to my ambitions had begun.

In 1947 I was invited to join the Isle of Thanet Specialist Club, now Kent Specialist Club, which had been formed for nomination from the National Flying Club and afterwards the British Berlin Flying Club. The reason for this was that our section was miles too large, extending to Southampton, thus we formed our own section in East Kent. Incidentally our NFC section to this day goes even further, to places like Reading, Aylesbury and beyond. The race points were Nantes and Bordeaux 1946 and 1947; three Guernseys in 1948; and then Brussels and Luxembourg in 1949. Under the guidance of Edgar Gowan, Bill Dray, Bill Fright and the one and only Stan Cecil, I started to learn my trade and to them I had the greatest respect. Their help with quality stock was gratefully received, much of which is in my bloodlines of my stock to this very day, especially the Bill Dray stock which was Delmotte Jurion with the annual late bred added from George Stubbs and Norman Southwell.

In 1950 I was called to do my National Service and left the birds in the capable hands of my mother who had helped so much financially, like a new Toulet clock for eight guineas. Whilst in the army I kept my interest in stock by keeping pigs in Greece, had a pet gazelle sleep on the bottom of my bed in Khartoum and the Red Sea Hills, and then kept and trained pigeons in Tel-El-Kibir on the edge of the Sahara Desert.

Back home in 1952 and it was 1954 before I was back flying to Bordeaux

Geoff Hunt holding 'Never Bend', 1st Barcelona BICC 1992, (684 miles).
Picture by Tony Bolton.

with the local club and Pau with National Flying Club, and then it all
began. W/Cdr W D Lea-Raynor, along with Frank Kightly, had persuaded
London Columbarian Society to organise a British section from
Barcelona. This International has always been the responsibility of the
Cureghem Centre which is based in Brussels and at this time was under

the leadership of the late Maurice Delbar as president. We were left to draw up our own rules but had to supply our distances in kilometres for the purpose of the International result. This is the case with International racing even to this day. Although any age bird can go, the organisers of the British section would only accept birds of three years old and upwards which had been verified in race time in at least one race of 400 miles or over. Through lack of support the race was discontinued in the early 1960s, so I had to be content with flying the National and Fed races until in 1975 when a British section was organised from Pau under the leadership of Ted Bennett of Hampshire 2-bird Club. In 1977 there was a Narbonne International with North East Kent 500-mile Championship Club, and so I was once again back in business.

It was a red chequer cock timed from the 1976 Pau International that was the start of one of my most successful lines to date. He is grandsire of 'Eureka', 1st Perpignan, 600 miles, with the British Barcelona Club. Pau and Narbonne brings me to the time the British International Championship Club was formed in 1978. I thought it kind of Frank Kightly at this time to offer the new club any help he could give, even to taking the birds to Belgium. At last my chance to achieve my dream and ambition seemed so much nearer but I had only just moved from Chilham to Westmarsh so I started once again to make haste slowly. Looking back I knew that I had taken many wrong turnings but my stock was pure Geoff Hunt and I learned that I had the right blood that could match all others, provided I produced my team fit, motivated and timed to the minute for any race entered.

The next three years I spent building a team so that they had experience and age on their side and in 1982 I began to feel that I was slowly climbing the ladder to success. In one day in June 1982 I timed four birds from Pau International, 559 miles, and four birds from Perpignan to win 1st at 600 miles. Eight in total in one day at the distance, but it didn't last for long as 1983 saw no overseas racing and my attention was turned to going to college for radio ham lessons. But each Friday afternoon at 3pm I'd phone the Ministry of Agriculture at Tolworth all through the summer and it was a case of "Who gives in first". I never did, and then on the Saturday it was down to see my MP, Jonathan Atkin, each week. It might not have done a lot of good but it made a good friend of Jonathan who likes a bit of marathon running and I still have a copy of one of his letters to the Minister telling him what a "cock up they were making".

Anyway, 1984 dawned with the overseas races back on the programme. This year was when for the second time I was 4th Barcelona and left still trying for the big one. In 1985 things got even better and I was 1st & 2nd Pau (559 miles), with 'Dusty Diamond' and 'Hermosa'; 2nd Lourdes (570 miles) with 'Gullito' and 2nd Marseille (586 miles) with 'Chigata' all with BICC international races. In 1986 I was 1st & 2nd Pau with BICC with 'Blue Diamond' and 'Oddy'. Good positions were taken in 1987 and then after trying on and off for 32 years I timed in my blue

'Sahall', winner of 1st Barcelona 1988, 29th Barcelona 1987, 10th Barcelona 1986, 6th Perpignan 1985.

chequer hen 'Sahall' (means Girl Friend) to win 1st Barcelona (684 miles) in 18 hours 39 mins, an ambition from 42 years ago.

'Sahall' had won 6th Perpignan in 1985, 10th Barcelona 1986, 29th Barcelona in 1987 and now 1st with a vel of 1076. Only seven days later the loft was 1st Dax (534 miles) with 'Vichte Princess' flying 14 hours 38 mins on the day with a vel of 1071 and winning 11th Open International Hens. I shall never forget 1988 as I had begun to think it was a case of not to be, but this win had put my name alongside the likes of George Stubbs and Record & Jallman who started the ball rolling back in 1931 and 1932 run at that time by Central Counties FC.

In 1990 I was 1st Pau with my blue white flight 'Hoist the Flag' and a week later was 3rd Barcelona with 'Micheco'. In 1991 I timed in 'Fair Trial' and a dark hen to win 3rd & 4th from Pau and then from Dax (534 miles) was 2nd with 'High Bid'. 1992 started in the usual way and I spent all my time planning for all five internationals, this being my first chance at all five as some members of my club had thought I should not take the birds to Belgium. I elected not to take the birds as this would allow me extra time to prepare a team for the last two races ie Marseille and Perpignan, most years I had not been sending to these two races. Talk about having the "last laugh", I started off with three home from four, winning 4th & 5th Pau with 'Fair Trial' and 'Sipsi Fach'.

Then I sent nine birds to Barcelona, liberated at 07.20 on Friday July 3 when the NFC were at Pau, Irish NFC were at Rennes and most other organisations held their birds because of continuous rain. I was left wondering if I was going to see any at all. About 9pm that evening Joseph Hackett rang to see if it was OK to fly over from Dublin to see the birds' home, so like the week before from Pau when Frank Durphy came over from Dublin, the pattern was set. Saturday dawned and it was a terrible day just as forecast. Joseph arrived and about an hour later 'Never Bend' was timed at 13.32 to be 1st Barcelona with a vel of 850 ypm with a two hour 26 mins margin to the next bird timed in. He gave me the Barcelona dream double and allowed me to join the history-makers like John Taylor in 1956 and 1957 with 'News Lad', C Wells in 1980 and 1981 and myself with 'Sahall' in 1988 and 'Never Bend' in 1992.

'Never Bend' is a grandson of 'Dusky Diamond', 'Hermosa' and 'Kefaak' who was 5th Barcelona to 'Sahall' in 1988. I ended up with eight from nine sent, some of which I hope will try again in 1993. At Bordeaux International I had 26 entered with 16 timed in race time and all home. Five more sent to Marseille to be 9th & 20th, all home and then to the final Open Perpignan; three were sent which had already flown Pau in 1992 and Bordeaux. 'Sipsi Fach' which had been 5th Pau was timed to be 3rd Open, 2nd BICC (600 miles), all three home to complete a very satisfactory season.

The lofts in use at the moment are made of wood and had all wire fronts until about three seasons ago when I decided to close the fronts in so that I could experiment by pairing the race birds up late, the middle of May. They are creosoted inside and out about every two to three years and the floors and nest boxes are done every year, with the nest box floors being done again a couple of weeks before pairing up.

Each loft has a nest box for each bird, ie two boxes for each pair. Hoppers are built in for grit and minerals. One loft is designed for flying open door, but the others have bolt wires with a landing board at waist height which I prefer. I find I can time in on most occasions direct from the landing board before entering the loft. Any sawdust which I use is first soaked with creosote and left to dry out before use.

The family of birds housed at the moment is related to some of the originals going back to the 1940s and over the years I have added top quality distance blood, mainly from National winners. Any named family means nothing to me as I have found over the years that by crossing and inbreeding it is certain lines that are most important and that nearly all my top birds over the years go back to one or two certain lines which keep cropping up time and again. By careful selection, testing and inbreeding I am able to find the one or two exceptional racers that I am always looking for, always remembering what has gone before will reappear later.

For feeding, training and racing I have kept the same system for many years, to keep altering is encouraging failure. It is best to make a convenient management system to suit your work and leisure and keep to it. The birds are hopper fed from March until the end of November with tic beans. That is every bird — racers, breeders and youngsters. The racers have a titbit early morning for the cocks and midday for the hens, and in the evening beginning the middle of May, they get maize fed by hand with the quantity being increased as the main distance events get nearer. The breeders get a small mix morning and evening, fed by hand into the nest box and the youngsters are fed wheat by hand night and morning with a little Red Band.

At the end of November the main feed for all is good quality feed barley (not malting) with a small amount of tic beans and a little maize if the weather is very cold. My day starts at 6am during the winter and between 4am and 5am during the racing season. I always listen to the shipping forecast and farming report and weather between 5.55 and 6.20am.

Most training is finished by 9am each day and I never train late in the day. As I only race the extreme distance races my method is as follows: The youngsters bred are not let out until near the end of old bird racing. They are then trained 30-60 miles the end of September. All yearlings and older birds go to one of the 236-260-mile races overseas the end of May beginning of June after which the yearlings are not raced. This allows me all of June and July to train and prepare the old birds for the long races. Every bird over the yearling stage has to go to 535 miles or further without exception. Training begins early May for every bird that flies out and goes to 100 to 150 miles, after which they are entered for a short overseas race.

I have three lofts which are paired at different times to suit certain races, the dates are decided so that each loft has a team of birds for races when they are sitting six-day youngsters. This means that most of the training is done while still separated. It is very important to me that the overseas race at 230-260 miles is one where they get a 7- to 9-hour fly, and if it happens to be very fast I would, as soon as possible, get them down the road again.

My team in 1992 went to Orleans (236 miles) with a midweek continental club, and were liberated in a strong NE wind. The early ones took 7½ hours and I well remember four landing together, flying 11¾ hours. All my team homed and went on to the five international races. I am certain they need this flying preparation as they have to fly all day and sometimes get up on the second day and do the same again. After this middle distance race my birds were treated with Sliepyard/Sliepsanol from Malibu Grains and given rue tea on Sundays and honey on Thursdays.

In the lead-up to basketing they were singled-up from 25 miles every day no matter the weather for about three weeks. In the case of my entries for Perpignan (600 miles), they had only flown around the loft since being timed from Pau six weeks earlier, so I took them overnight by car to my friend Brian Howe of Dorchester and he kindly singled them up for me next day (177 miles) flying into a NE wind. This was nine days before the race and I finished 3rd Open, 2nd BICC with all three home.

I would like to offer some advice to anyone who is thinking of specialising in the extreme races. First you need to obtain the right blood. The best to my mind is still the old English families that are still very much individual ones and not follow-the-leader types. Obtain this blood from someone with the stock that is performing at the present time. Only buy a few, say six or eight, and put yourself in his hands and keep to a system the family are used to. If things don't go as expected, usually only a small alteration in the management is necessary. Distance birds don't need to be raced every week so try not to get involved in racing week in and week out. I have found that my best birds are bred from ones that have performed well when flying in east winds, I try to pair these together.

Always try to keep things simple. The answer to success is not in a bottle but goes to those who have motivated and timed the condition of their entry to the very last minute, I never make my final decision for entries until an hour before basketing. Always use commonsense and never be afraid to seek advice from fanciers who race these kind of distances. Don't despise and discard a pigeon that has its heart in the right place, just remember looks are only skin deep, and above all don't be discouraged by failure.

I have never been happy with the way our birds are transported by road. Maybe it is because I have seen the way our international birds are convoyed to the race points. Going by train with tons of space and water at all times, the returns from these races are very good and their condition is excellent. I would like to see the crates on our transporters oblong in shape, water on three sides and more space between crates plus of course more between top crate and the roof. Now it is more evident that less time is being taken between basketing and releasing our birds in the distance races. Birds in transit need watering every four to five hours and need a full day's rest at the race point with, as the old fanciers used to say: "The sun on the basket", and I know there is a lot in what they said.

continued at foot of next page

PARAMYXOVIRUS VACCINATION — THE FACTS

by A S Wallis BVSc MRCVS

The threat of paramyxovirus became apparent from southern Europe and Africa during the early 1980s but it was only in June 1983 that it appeared in a loft in St Austell, suggestively, one in which stray birds with Portuguese rings had appeared. The symptoms seen in the affected lofts were very reminiscent of Newcastle disease in chickens and tests carried out on the loft owner's kitchen table, confirmed that the birds were infected with a virus of the same family, family I stress, not identical, similar to but there is a lot of variation within the group of viruses which are associated with Newcastle disease anyway. When the disease eventually spread to chickens by indirect and unusual means, the signs were recognisably those of Newcastle disease but with some features not previously encountered. All of these viruses are included in the paramyxovirus group 1, quite a varied mixture though the pigeon viruses are quite distinct within it.

Over the years, control of Newcastle disease in poultry by vaccination has proved to be the most stable and efficient method and this led logically to the idea of vaccination of pigeons. Tests were undertaken using existing chicken vaccines, two of which were found to be effective and became licenced for use in pigeons from September 1983 onwards. Since that time other vaccines have come onto the market and are of course fully licenced as safe and effective in pigeons.

These more recent comers were not chicken vaccines but vaccines adapted for use in pigeons. All of the current vaccines are 'dead' that is to say, they are based upon quantities of paramyxovirus type 1 which have been completely destroyed and which, therefore, are in themselves

---

Continued from previous page

Looking into the future my new ambition is to race from San Sebastian and, of course, to win an International outright. Wishful thinking maybe but we shall see. In closing this article on the years building up to the "Barcelona Double" I just hope it has been of some interest and don't forget to ask if you have any queries.

I would like to say thank you to all the lads at the Wingham Flying Club and Malibu Stud for their help with setting and checking my clock and to my wife Annie and son Ian, for even living with me and answering the phone for months on end. To everyone I say: 'Have patience and make haste slowly" and just remember, the laurels will always go to those with the imagination to win and with birds that are well prepared and supremely fit and blessed with that incalculable ingredient — the willpower to keep going.

quite incapable of giving rise to the disease itself, though are nonetheless capable of stimulating the pigeon's immunity organs to produce disease resistance.

Chicken Newcastle disease vaccines are of two types, killed vaccines similar to those used for pigeons and also living vaccines. These latter consist of preparations of living virus either bred for the purpose or discovered in nature which whilst capable of growing within the chicken and providing a sufficient quantity of virus to stimulate the development of immunity within the chicken are nonetheless, virtually unable to produce disease at all. From time to time, circumstances may arise in which secondary disease appears due to bacterial invasion which can be quite destructive in itself. There is only one live Newcastle disease vaccine strain licenced in the UK — the Hitchner-B$_1$ strain.

Dead vaccines have to be administered to each bird individually, usually by injection and the dose volume is quite large (usually 0.25-0.5ml) since each dose must contain all of the virus material required to stimulate immunity. So far as live vaccines are concerned, these are intended to grow within the host to produce a sufficient quantity of virus to achieve their effect in stimulating immunity so that only a small seed of virus is needed to be introduced into the vaccinated bird. Consequently, doses are very small indeed and the vaccines can be given by mass vaccination methods in the drinking water or as a spray or indeed individually as a drop in the eye. Live vaccines are relatively cheap, a bottle containing 1,000 doses generally being bought for less than £5.00.

Live vaccines also work in a rather different way from dead ones in that although the immunity to which they give rise is not so intense or long-lasting, it does develop more quickly than that derived from dead vaccines. In the poultry industry both types of vaccine are given as part of a vaccination programme: live vaccine early in life for cheap and rapid protection, dead vaccine later on for solid, long-term protection of breeders and layers.

So how does all of this relate to pigeons?

Dead vaccines are very effective in the basic programme for the protection of the loft against paramyxovirus. The older vaccine required two doses to be given but those more recently introduced give perfectly satisfactory protection with only a single injection and this will prove sufficient for about one year — providing that the birds were correctly vaccinated at the right age.

Immunity though is not an automatic machine-made thing to be switched on and off at any particular time. Once it has developed, immunity wanes slowly over a period of time. It must be considered along with the risk so that, where there is paramyxovirus in the district and the loft was last vaccinated perhaps ten months ago, one would be wise to re-vaccinate early rather than go for the letter of the date since more recently vaccinated birds have a much stronger resistance than those vaccinated many months before.

So what about live vaccine? Hitchner-B$_1$ vaccine (and its hotter cousin La Sota not licenced in the UK) are not very effective in pigeons because they are derived from typical poultry strains of paramyxovirus which are not well adapted to the pigeon and will not lead to the sustained virus multiplication required for a solid immunity. Hitchner-B$_1$ vaccine does give rise to a measurable immunity in pigeons, however, it is not very strong and it does not last for more than a couple of weeks or so but, such as it is, it does develop rapidly within five days whereas dead vaccines take twice that time to produce any measurable response.

Given an outbreak of disease in a unvaccinated loft, therefore, there is a good case for using both live and dead vaccines together to get the benefit of the early response from the live but needing the weight and strength of the dead to finish the job. There are just a couple of points, live vaccine is only licenced for poultry not for pigeons and therefore may only be used in pigeons or supplied on prescription from a veterinary surgeon and also live vaccines can infect people, giving rise to a very sore conjunctivitis and cold-like symptoms.

Will there ever be live vaccine developed specifically for pigeons? On the whole I doubt it. The cost of development would be relatively high for what is a fairly small market (the development costs of Hitchner vaccine were written off probably 30 years ago) and since live vaccines tend to give fairly short-term protection anyway, it is difficult to see a situation in which a single live vaccination would give rise to a year of useful protection. In the end, the greater duration and predictability of immunity resulting from dead vaccine is likely to carry the day and whilst more expensive, taken against the cost of one's hobby over a year, it really is not so heavy a consideration.

The safety of vaccination procedures has long been the subject of scrutiny. One particular brand of dead vaccine in the early days was associated with severe local reactions and the growth of large lumps on the vaccinated birds but this brand has long been withdrawn. Occasional sudden deaths have been reported from time to time. The cause of these seems to be mechanical interference with the complicated and very rich network of veins in the neck region of the pigeon and provided that this is avoided carefully by directing the injecting needle backwards away from the neck rather than forwards into it, these problems should not arise and indeed, it is a long time since the author was last consulted about this problem.

Over the years, a number of other problems, empty eggs, poor returns to name but two, have been laid at vaccination's door. I have investigated some of these, but as problems they were common enough long before paramyxovirus appeared and no solid evidence has ever been produced to me which incriminates vaccine or vaccination in this respect. So far as Hitchner-B$_1$ live vaccine is concerned, its lack of effectiveness in pigeons alone suggests that it would be very safe, since it does not multiply significantly within them, probably a lot safer for the pigeons than perhaps

continued at foot of next page

DREAM YEAR

by Don Wilcox of Wilcox & Evans, Talywain
1st Welsh National Tours, 3rd Welsh National Thurso

The year 1992 has been one of those seasons of which people dream, and the ultimate achievement was to win 3rd Welsh Grand National, 1st East Sect from Thurso, winning a sponsored car and an STB clock. Other wins at National level achieved this season are as follows:

Welsh South Road NFC: Tours 1st National, 1st East Sect; Pau 12th National, 2nd East Sect and 20th National, 7th East Sect; Nantes 12th National, 1st East Sect; Rennes 16th National, 2nd East Sect. Welsh Grand National FC: Thurso 3rd National, 1st East Sect, winner of car, winner of STB clock; Crieff yearling race 17th National, 2nd East Sect. New North Fed Open 1992: Crieff 4th Open; Thurso 1st Open. Total prize and pool money of over £1,600. The cock that won the sponsored car has been named 'Dream Year'.

I first started racing pigeons in 1955 with my uncle, who, unfortunately, died after two years. However, in those two years I probably learned as much about pigeons as would take another novice five years. My father than joined me in partnership until he died in 1973. Mike Evans, my present partner, joined me in 1973 and we flew as Wilcox & Evans for approximately four years, until Mike had to finish due to pressure of work. However, in 1985, we re-formed the partnership and we have flown together ever since. Although we are totally of different personalities, we find that we have as near a perfect partnership as one would hope to achieve.

Since 1986 we have achieved the following positions in National races flying with the Welsh South Road NFC and with the Welsh Grand National FC and Welsh Combine. Such wins as: 1st & 28th Open Pau (only 31 in race time); 3rd Open Nantes; 4th Open Guernsey (YB); 6th Open Guernsey (OB); 9th Open Nantes; 2nd & 6th Open Pau; 2nd & 16th Open Guernsey (YB); 16th Open Rennes; 12th Open Nantes; 1st Open Tours; 12th, 20th & 41st Open Pau (only four sent); 14th, 15th, 16th, 40th &

Continued from previous page

for the vaccinator.

The control of infectious disease by vaccination has a very long history and is perhaps the single most effective disease control procedure proposed in the history of medicine. Nothing in life is completely perfect but it has been realised for 200 years that the inconveniences of vaccination far outweigh the tragedies of infectious disease. There is no reason to think that this is any less true of the control of pigeon paramyxovirus than it was of human smallpox.

Some of the trophies won in Open and National competition by Wilcox & Evans of Talywain.

44th Open Littlehampton; 13th Open Nantes; 6th & 7th Open Nantes; 12th Open Pau; 6th Open Cherbourg; 10th Open Nantes; 6th Open Pau. All these plus scores more prizes in first 100 Open National since 1987 alone.

On the North Road our wins since 1986 are: 1986 — 4th Sect (2,538 birds), 4th Open Welsh Combine New Pitsligo (7,049 birds); 1988 — 4th Sect, 25th Open Thurso WGNFC (3,233 birds); 8th Sect, 27th Open Lerwick WGNFC; 23rd & 44th Sect, 51st Open Carlisle WGNFC (4,764 birds); 4th, 13th & 30th Sect, 8th, 27th & 74th Open Roslin Park YB WGNFC (2,825 birds). 1989 — 2nd Sect, 12th Open Thurso WGNFC; 8th Sect, 18th Open Welsh Combine (9,027 birds); 4th Sect, 11th Open Lerwick WGNFC; 5th Sect, 11th Open Welsh Combine. 1990 — 35th Open Thurso WGNFC. 1991 — 9th, 39th & 40th Sect, 26th Open Thurso WGNFC (4,199 birds); 17th Sect (2,585 birds) Welsh Combine; 1st & 16th East Sect (1,537 birds), 1st Combine (2,814 birds) Newton Grange (290 miles). 1992 — 2nd Sect, 17th Open (3,746 birds) Crieff WGNFC (Yearling); 1st Sect, 3rd Open (3,337 birds) Thurso WGNFC plus motor car and STB clock.

We fly the Natural system old birds and we race approximately 17 pairs North and 17 pairs South. The birds are let out in the morning about 7 o'clock and have an open hole and they are out until dusk. They are trained about three or four times per week, 30- or 40-mile tosses. Feeding in the morning is a mixture of Haith's Red Band and crushed maize and

continued at foot of next page

CONCORDES NOT JUMBO JETS

by Dr G A Richmond
1st Open Midland CC Fraserburgh

It gives me great pleasure to write this article for Squills, in part because it is a tribute to my winning pigeon which without doubt is the finest bird I have ever owned. His talent suggested itself when as a YB in 1990 he was 10th Sect A NRCC — with 1,169 birds in the section, but as a yearling he was only consistent, being totally outclassed by his brother. This year however 'Wazo', as he has been christened, has excelled himself. His first win was 9th Open NRCC Perth (5,707 birds) winning a cash total of £333, and £100 Louella voucher as runner-up to NRCC Perth car winner. The following week from the same race point, 'Wazo' was 7th Open winning £507 in Midland Championship Club. With the same club from Fraserburgh he was 1st Open winning £925 and a Gold Medal.

Although I have had numerous Fed successes over the years, including two Gold Medals, the wider challenge of NRCC and MCC racing is a truer and more difficult test of both fancier and pigeon. Such racing demands an intelligent bird and although the joys of success can be great, the disappointments can also be extreme.

Many will wonder how a doctor began racing pigeons. Most people can think back to a moment in their life when a single event changed their lives. With me it was the moment when our family cat caught an exhausted racing pigeon. This bird was eventually to die, despite our care

Continued from previous page

wheat. The cocks are fed about 10.30am, the hens about 2pm, and then they have their main meal about 7.30 in the evening, which is Willsbridge No 2, a mixture of beans, peas, tares and maize.

The loft is cleaned out three times a day, and the water is changed three times a day. Grit is fed by hand at each feed time. Always in the loft is a pot of Kilpatrick minerals. Cod liver oil capsules are administered about once a week, and PLG is given to them after training and racing as a preventative against diseases.

For the novice I suggest they acquire at least 30/40 youngsters to start the season. Race them all to about 100 miles and then stop half the team; the remainder can then race to the end of the programme. Hopefully, by this means they should have a sound team of yearlings for the following season. I suggest that they purchase pigeons from a good local fancier. Feeding methods are such that the birds are fed the same time regularly every day, given plenty of clean fresh water, and have a clean loft. I would like to thank the Editor for inviting us to contribute this article for the 1993 Squills Year Book.

and attention, but the fascination by that time was complete. The death of my grandmother soon after meant my father had sufficient funds to put together a loft which he did with large metal brackets and dozens of bolts! Local fanciers soon rallied round with surplus birds and the fascination blossomed. Later I went into partnership with the late George Ingham of Shelton Lock who needed a younger pair of legs to keep his sport alive. Unfortunately the death of George some two years later and my imminent 'A' level exams meant the end of pigeons. After University and a medical degree I settled into general practice in Alfreton, Derbyshire. Within no time at all, contact with the then declining but active mining community rekindled my enthusiasm as loft after loft presented themselves on my daily visits.

In 1970 my own loft was erected and stocked, at no cost, by my good friend the late John Stubbins of Creswell. In John I found the perfect friend and adviser to take me forward into the serious competition of the area. Indeed in my first race I was 3rd with a January-bred young bird from John and this same bird topped the Fed in my second race with a horrible velocity of, from memory, about 880 ypm. This team won the YB Averages with ease and on this firm foundation future successes followed. The second of my Fed Gold Medals was won by a bird again bred by John. Involvement in administration soon followed when I became president of Notts & Derby Border Fed and secretary of Midland Social Circle. The presidency of the Fed still takes up considerable time and thought but the Midland Social Circle job is now in younger hands.

Indeed 1992 was an excellent season for me, with other birds apart from 'Wazo' performing well. However, Thurso the next MCC race was different. I began writing this article at 5.30am waiting for an early morning second day bird! The dawn was pleasant, with the air cool and light skies, although the sun was not yet above the horizon. With MCC, which is now a two-bird specialist club, continued success is difficult. As time passed it became obvious that without my ace 'Wazo', life was more difficult. There had been 18-day verifications out of the 134 lofts and initially my optimism had been high. As the sun rose and my second cup of black coffee vanished it was obvious my wait was in vain. Gradually all enthusiasm for the race disappeared and my Fraserburgh success seemed so distant.

Perhaps it is the uncertainty of any result which is part of the fascination. Especially in NRCC racing where now the sections are much more even numerically, birds will gain positions in the Open very often against the predictions made by the wind on liberation. On good racing days in club competition, however, everything is very predictable. In my own club which sends an average of 300 birds weekly there are several keen experienced Widowhood fliers. On good racing days the first four prizes have disappeared within ten seconds and the winning lofts are highly predictable.

One obvious truth is that if you don't compete you will never win, and to win you must risk the possibility of also coming last! Pigeon racing must be one of the greatest levellers of all time. Yet as manager of your pigeons you can do much to minimise the risk of defeat. It is important that pigeons are healthy, trained to a level of appropriate fitness, not exhaustion, and experienced. I do not mind if my young birds have difficult races as long as they come home and appear to learn from their mistakes. One of my best birds which was 2nd Open MCC Lerwick with only five day birds, was lost as a YB. This hen returned in the March of the following season and once down on eggs competed in several races to Berwick. The next season she had several club prizes before the Lerwick success. Birds with this history very often make excellent birds in the future if given the chance.

Having now experienced the pleasure and success of Widowhood, it is obvious that this system has a great deal going for it. Nevertheless I am equally certain that a team of Natural pigeons is equally important for the distance races. More distance events have been won by hens than cocks and these days hens usually make up a smaller percentage of the convoy. Certainly they seem to have more stamina than cocks and excel especially when the racing is hard. In such conditions Widowhood cocks do not carry sufficient weight for the marathon ahead and have actually lost before they are liberated.

Good stock is best obtained from successful local fanciers who themselves are ruthless in their standards and selection. Fancy foreign breeds may well win but the likelihood of success with birds from an established racing loft must be higher. Success needs both patience and ruthlessness combined with thoughtful observation. My best birds are well balanced, well featured and feel calm and comfortable in the hand. They have a bright and healthy eye (not eyesign), but above all show some individuality and character. It is true that birds come in many shapes and sizes but to my mind a medium-sized, well-balanced bird which does not carry great weight is the right type of bird to aim for.

John Stubbins used to say we are aiming to fly Concordes not Jumbo Jets! Although I have had some good Jumbo Jets in my time, 'Wazo' is definitely a Concorde — his winning Fraserburgh velocity — 2078 ypm.

MOTIVATION

by Sam & Rolly Fear, Clandown
1st Open Pau Central and Southern Counties FC

It was a pleasant surprise for my brother Rowland and myself to receive a call from the editor of Squills and it was thanks to our blue hen GB90F14401 which was bred from our old family of birds named Motivation. After our great success in the 1970s we had become pigeon keepers rather than pigeon racers. I myself had become quite happy to motor along winning now and then, but our many friends in the Fancy were always offering us birds to get us back again. Many thanks to them.

At marking one evening at our club HQ we were having a "jar" when I was asked what I thought was the "most important things for the success of pigeon racers". I answered "motivation" of both fanciers and pigeons. This set me thinking. I decided to look up the record books of the birds and eliminated quite a few from the loft, but that's another story. Next we had to consider their health so I cleaned the lofts out with disinfectant and blowlamp, then we decided on a mixture of beans, peas, tares, maize etc, for their diet.

We mate our stock birds one loft February 1 and the Natural loft the middle of March to the middle of April. We have flown Widowhood but Natural is the game for us, with our own innovations. We play every day and week by ear taking into account the weather for their diet etc. We like to separate the birds sitting from 10-16 days for a distant race of 500 miles plus.

With the young birds we fly them Natural to their perch, only letting them out in the evenings until we start to train them. We would like to train every day 15-25 miles if possible. They are fed a mixture and a few pellets with linseed and hemp half-and-half, the only seed we feed.

Our birds originally came from Bert Thatcher of Radford, Nr Bath. They were winning Thurso north 500 miles plus Pau south 550 miles, and line bred, and looked like peas in a pod on the perch. Bob Legg of Seabrought, Nr Crewkerne, a farmer who never had much time to race his birds bred us a few youngsters to race for a couple of years. We still have the bloodlines today flying "distant Natural". When trying out Widowhood we exchanged birds with Swain Bros, Dorchester, and they did very well for us and won.

For the young fancier just starting I would say get a good dry loft; go local for your stock. Visit one or two good fanciers who are winning regularly then study the diet for those you want to fly — Widowhood or Natural. Find out and read everything you can. Ask other fanciers if you can visit their lofts. Most fanciers will be only too pleased to show you their lofts, and their birds, at the right time.

Visit the shows in the winter to see how birds look in condition. We

are still learning about pigeons, there are always a few surprises. Hoping these few lines will interest a few fanciers, wishing you all the best of racing and showing for the coming year.

CHAMPIONS OF SCOTLAND
1992 RENNES WINNER

The small loft with the big performances
by John Honeyman & son, Kennoway

My father, James, along with his three brothers raced pigeons for as long as I can remember. Naturally I took to the sport and joined the boys' club in Kennoway. Progressing to the adult club I aimed to be as successful as my father and uncles. Being a member of Kennoway HS for 25 years, I gained numerous positions within the club and Fife Fed. About ten years ago I joined the Scottish National Flying Club and thoroughly enjoyed the long distance racing. Like all National fanciers my lifelong ambition has been to win the country's most coveted, prestigeous prize — the Rennes race.

I fly as J Honeyman & son with 12 year old son John Junior — and my pigeon loft is in my back garden. It is not a glamorous, fancy or expensive loft, only a basic 14ft long loft with two compartments. "The small loft with the big performances". There is no extra room for expansion as I live on a small housing estate with a small garden. I never neglect my pigeons even for a day and am very strict about cleanliness

John Honeyman of Kennoway, Fife, 1st Open SNFC Rennes 1992.

J Honeyman & son holding 'Honeyman's Dream', 1992 Rennes winner Scottish National Flying Club.

in the loft with clean, fresh water, changed twice daily. On eavesdropping at the marking stations I quickly realised that good 500-mile plus stock birds were needed to found my loft. I sought out, and purchased pigeons from fanciers who were very much to the fore, and being patient I have built up a family of birds, over three to four years. My birds are entered for the club races and certain pigeons fly the full programme within the Fed. These races are essential and good training for the birds which are intended for the Channel races.

Success in the Channel races led to a Gold Award certificate in 1989. This is given for consistent success in long distance races, ie for a bird with five Open position certificates in Channel races. I was very proud to gain this award and it made me strive to do even better. Having only 15 pairs of old birds I selected two for this year's Rennes race. Both were gleaming, with good bodies and very keen on the specific nest conditions I had set up for the birds. Every Scottish fancier dreams of winning the Rennes race, the Blue Riband race of the National year. On winning the race and realising my dream, I aptly named the winner 'Honeyman's Dream'. The cock is only two years old and is bred from 'Fallowhill Superstar', a Rennes winner in 1986. The 'Dream' is retired from racing, for stock purposes, of course! For the last Channel race this year Sartilly (2) I selected another two birds for the race, again set up in the nest conditions to which they prefer. I gained 18th & 173rd Open positions. Two birds sent, two birds winners. The latter position resulted in receiving my second Gold Award certificate. I am truly delighted, even estatic, about my performances this year. All the time, hard work and training has paid off.

'HONEYMAN'S DREAM'
First Open SNFC Rennes 1992 winning Gold Cup, King George V Cup,
H A French Rose Bowl, British Homing World Trophy, Pacemaker Trophy,
Flying Festival Trophy, East Sect Trophy, 1st Fife Bird, J McLeod Memorial
Trophy, 1st Club Rennes Trophy, 4,501 birds, vel 1163 on the day, 563
miles.

My pigeons fly on the Natural system, the same system that my father
had used. Birds are dewormed and treated for canker prior to breeding.
Health grit and mineral blocks are available to them all the time. A special
titbit of Red Band is given midweek during the racing season, I never
do any alterations during the racing season, I feed at the same time each
day. I truly believe that many fanciers' problems result from overcrowding.
Strive to avoid this, it is easy to keep too many birds. Good pigeons are
essential, only pair good hens with good cocks. Introducing too many
strains of birds is another cause of the downfall of many fanciers. Build
a family of birds which are all related.

Fortunately I have never had to consult a veterinary surgeon, as there
has never been any serious diseases in my loft. Keeping a keen watchful

'SPENCE SPECIAL'
Gold Award winner 1989. SNFC results: 1985 1st Club, 43rd East Sect,
52nd Open Sartilly (2) 3,395 birds; 1987 1st Club, 86th East Sect Sartilly
(1) 2,358 birds; 1988 1st Club, 43rd East Sect, 54th Open Sartilly (1)
2,883 birds; 1988 42nd East Sect, 61st Open Sartilly (2) 1,929 birds;
1989 2nd Club, 105th East Sect, 159th Open Sartilly (1) 2,989 birds.

eye on the birds, observing them, results in noticing any ailments quickly
before they progress. I then act accordingly. Note that medicine is only
given when necessary, not given as a precaution. My birds are fed on
Knock-out mixture and conditioner supplied from J W Swainston & son,
Durham. There is a local stockist in my area. Avoiding dampness in the
loft at all times is of the utmost importance.

Fanciers should look for consistency from inland races and keenness
prior to Channel races. If the racers do not exercise themselves then
training is increased. Pigeons must be set for specific races. Experiment
with different nest conditions and then observe how the pigeons react.
This requires time and patience but has paid dividends for me. I am my
own man so I do not listen to others but do my own thinking. Like any
good fancier I am now planning and dreaming of 1993 racing season.

'GOLDEN DREAM'
Red pied hen, Gold Award winner 1992. SNFC results: 1986 1st Club, 6th East Sect, 12th Open, 1st Fife Bird Sartilly (1) 2,582 birds; 1987 22nd East Sect, 44th Open Sartilly (2) 2,208 birds; 1988 1st Club Sartilly (2) 1,929 birds; 1st Club, 11th East Sect, 12th Open, 2nd Fife Bird; 1991 48th East Sect, 98th Open Sartilly (2) 3,673 birds; 1992 1st Club, 98th East Sect, 173rd Open Sartilly (2) 2,915 birds.

Being 40 years old I hope to have many more years to enjoy my hobby and sport, and perhaps "do the double"! After all no-one ever imagined that the Rennes winner would be in Fife, as the birds fly 563 miles. The record of 36 years has now been broken.

A FAMILY AFFAIR

by Kevin Murphy of J Murphy & son
Winners of 1st Scottish NFC Sartilly (1)

First I would like to thank the Editor for inviting me to write an article for Squills, it is indeed a very great honour. Secondly I would like to dedicate this article to my partner — my dad, Joe. Never have I met anyone who is so dedicated to our sport and to whom I have learned everything from. Thanks dad.

My dad always says that if you put enough work into anything it will eventually pay off. That was our secret — hard work! Owing to the fact both of us were off work due to ill health, we were both two full-time doo men, both with different views.

I have been interested in the birds since I can remember. My grand-dad kept them and my dad started in 1970 when a stray had come in and was put in the garden shed, my mum was caught in the shed feeding and speaking to it. Needless to say the shed was soon converted into the pigeon loft.

Our most important aspect of pigeon management is hygiene. Our lofts are scrubbed out every Sunday morning one week with Virkon S from Vetrepharm, the next with bleach. We have good ventilation in the loft, therefore, the loft dries very quickly. We hate to see pigeons in dirty conditions.

The most important thing I think about National racing is conditioning the bird. It doesn't matter how well bred it is, or how much it cost, if it isn't fully conditioned it will never win. Many people send birds 500-600 miles, few can condition them.

Having made alterations to the hut, a new roof, renewing the outside but trying to keep the inside as unchanged as possible, and improving the ventilation with the help of my friends and dad's mates during the winter, it was now down to us.

Although my dad wasn't working the birds were still cleaned out at 5am every morning. Every weekend the floor and nest boxes scrubbed. A pigeon won't be healthy if it's picking dirt and disease from inside the loft itself. In previous years I would look after the young birds and dad doing the old birds. This year we decided we'd work together as a team.

When dad started back to work, the ground work was all laid. Dad was up at 5am cleaning and fixing the old birds before going off to work. I was up at 6.30am flagging the racing team as dad went. At dinner time the hens were given the open bowl as the hut was cleaned out and water topped up.

At 4pm the racing team were again flagged as the hut was cleaned out again and water refreshed using Vetrepharm products as for two years. The racing team were kept going until dad came home from work. When

Kevin Murphy holding 'Mystical Rose', 1st Open SNFC Sartilly 1992.

he left for work I was standing flagging them and when he home he found I was still flagging them. I think he thought I stood there all day.

Our winner a four year old chequer hen 'Mystical Rose', named by my mum, was bred from a Victor Loch cock and a John Bosworth hen. Previously she'd won 1st Club, only bird on the day, 8th Open Fife Fed Sartilly, 16¾ hours on the wing as a yearling; as a two year old flew Rennes; as a three year old, 1st Club, again only bird on day, 1st Fed, 1st Region C, 20th Sect E, 83rd Open SNFC Sartilly (1), 15 hours on wing; five weeks later 144th East Sect, 236th Open SNFC Sartilly (2). Then finally this year 1st Club, 1st Fed, 1st Region C, 1st Sect E, 1st Open SNFC Sartilly (1) 3,293 birds, also winning the Ogilvie Trophy for best individual performance for SHU.

'Mystical Rose' tells you herself when she is right. You just have to look at her. She goes a lot darker in colour, the same thing happened at the last SNFC Sartilly. A young late bred cock sitting on the perch, he was outstanding. This late bred was sent with two nest flights to only his fourth race of his life and was timed after 12½ hours on the wing from 515 miles to win 36th Open 2,915 birds.

About three weeks prior to winning the National we noticed that 'Mystical Rose's' cock wasn't looking too well, so dad took him out of the loft. He died during the night; we think he'd been poisoned. 'Mystical Rose' was paired to this cock for two years and wouldn't look at another. After trying four different cocks with her (she battered all the previous ones) a young cock was showing off to her, so at night he was put in her box. She laid right on time sitting her favourite condition, 15-day eggs.

We sent seven birds to the Sartilly (1) with three on the result, 1st Open, 141st Open, 127th ES and had six out of seven by the next morning. Next came Rennes, sent six birds timed two on the day. Our first pigeon 'Rising Star' was 39th Sect E, 53rd Open, 15 hours on wing; second was 'Matt's Mealy' 55th Sect E, 75th Open, 15¼ hours on the wing, after failing to time out of Nantes. Then the final SNFC race from Sartilly, sent six, five returning home. First timed in was 'Lady Janet' winning 1st Club, 1st Fed, 1st Region C, 4th Sect E, 5th Open. Our second bird was 'Nick Faldo' which was the late bred cock which is mentioned earlier which won 36th Open SNFC Sartilly (2).

I must admit to being totally over the moon on winning the SNFC Sartilly race, and anybody who saw me that night will surely agree. It was more satisfying than winning one hundred 1sts in our local club. The hard work had paid off.

Before I end I would again like to thank the Editor for asking me to write this article and also to everyone who congrulated us on our win. It was a lot easier getting up at 6.30am to flag the birds. But I personally would like to say a special thank you to my mum, Margaret, who puts up with a lot, especially from dad and I on our "different views", but also to dad to whom I am deeply indebted, for without his help, dedication and constant nagging I would never had been so proud to be his partner in winning 1st Open Sartilly. Maybe one day dad, I can return the favour. Finally wishing all our friends and everyone else in the sport every success you deserve in 1993.

THE FAMILY THAT WINS COMBINES

by Steve White of S White & Sons, Dagenham
9th, 3rd, 2nd Combine, yet beaten for the Tommy Long Average Trophy

I'd like to begin my story by thanking The Racing Pigeon for putting my story into Squills Year Book. When I was single and still lived at home my father raced pigeons, I always kept and bred cage birds and finches, went out poaching etc. I met Jacki in 1980, and we moved into a trailer (mobile home) at Noak Hill, Essex, where I kept my cage birds. In 1986 we got a house in Dagenham, and I decided to try my hand at racing pigeons, as Jacki also comes from a family that race pigeons.

The house we moved into had rather a small back garden, 16ft x 36ft, so I had to arrange my lofts as best I could, across the bottom, one half for stock, and in the other half I raced YBs. My first season was YB 1987 in the Crowland FC. I raced a small team of approx 24 YBs. My stock birds were from my father, father-in-law, and my brother. Mainly Haeltermans. In my first season YB 1987 I lost my novice status on the third race. That season I won 14 prizes in the first six. After the 1987 season I decided I really wanted to achieve even more, so in 1988 I first tried the Widowhood system, and I won 22 positions in the first six. I'd changed the loft to an L-shape, 16ft across the bottom and 9ft x 4ft shed up the YB side of the garden. 1988 was the first year to get any Combine cards, 36th Open Berwick Yearling OB; YB 13th, 68th Open Morpeth, 5,843 birds. This Combine was won by my brother, I bred his Combine winner in my loft. I'm one of three people in our family to win a Combine; my father won 1981 Morpeth Combine, my brother won 1987 Morpeth, and myself a double Combine 1991 Berwick/Berwick Yearling. In 1988 in the Crowlands FC I was second highest prize-winner, winning Scottish Av, Combined Av, YB Av and Thurso winner.

During the winter of 1988 I decided to change the lofts around, I tiled the roof, which is apex style, installed heaters and ionisers, I built hen boxes for the Widowhood hens, which were kept in a small shed, I also glazed the windows with glass, where I'd had plastic, so I was ready for the 1989 season. That season I got off with a flying start, and took sixteen 1sts, seventeen 2nds, thirteen 3rds, twelve 4ths etc. In the LNRC Berwick Yearling I was 3rd, 5th, 11th, 30th; Berwick 47th; Stonehaven 32nd, 43rd, 57th; Thurso 25th, 27th, 79th; Morpeth 13th; The Combine averages 9,000 birds plus. The trophies won this season were: Highest prize-winner OB Av, YB Av, Scottish Av, runner-up YB Combined Av, Best Av Berwick, Morpeth, Thurso Cup, Best Av over four Combine races, LNRC Trophy, the Up North Friendship Trophy for runner-up Best Av over four Combine races, pools and prize money that season were over £2,000.

After this season I knew I had the right birds, I was really pleased with the way they worked on the Widowhood system. 1990 season was even better. My birds, or as I call them my family, were working even better for me. I took eighteen 1sts, thirteen 2nds, ten 3rds, seventeen 4ths, fourteen 5ths, seventeen 6ths, and most of all the averages I'd won the previous years. My Combine positions were 7th, 9th, 52nd, 52nd, 60th, 90th, 90th. This was the first season I'd won the 2-bird Championship Club, which I was really pleased about, 1st Gateshead, 1st Stockton, prize money £2,000 plus pools.

The loft was working for me, so I didn't change a thing after these few seasons of racing. Most of the comments I'd hear were: "He won't keep it up for long", but unbeknown to me 1991 was my golden year — two gold medals. I won Berwick/Berwick Yearling Combine with 10,930 birds. I collected an RPRA award, forty-one 1sts, twenty-one 2nds, eighteen 3rds, twenty-four 4ths, nineteen 5ths, thirty 6ths, flying in three clubs and three Feds. This was my best ever year. I won most club's and Fed averages, I also won many more Combine positions. The 1992 season I started doing well, but my birds didn't seem right, edgy, nervous, something wasn't right, and then the answer. One night Jacki was checking on my eldest son Bobby, she noticed one or two sensor lights kept coming on. Jacki looked out of the window and there was the problem; a mother and six fox cubs in next door's back garden. When I checked the next day, the fox had dug in under my loft. The noise of the foxes during the night was disturbing my birds. I even set up the baby's monitor in the loft, to see how loud the noise was. The fox had even dug under the loft in my other neighbour's garden. I have a Staffordshire Bull Terrier that lives in the back garden, but she couldn't get to them, so all I could do was cement in the holes and break up a load of glass, which seems to have worked so far. I've had 30 wins so far this year, among many Combine positions this year I managed 9th Berwick, 3rd Stonehaven, 2nd Thurso and runner-up Tommy Long Trophy, so close once again (I'm going after it next year Tel!). I can't say much about the rest of this season yet as it hasn't ended at the time of writing.

Widowhood cocks are paired up January 20, when the chicks are 12 days old, the hen and one chick are taken away, leaving the cocks one chick to rear. My cocks are given a couple of short training tosses, while bringing up the chicks, when the chick is old enough to take away the cock is then put on the Widowhood system and is only shown the hen when they come back from racing. I have found on the Widowhood system that my birds can fly up to Thurso, 508 miles. I only show the bowl Friday night on basketing for 30 minutes. I like to keep the temperature in the loft about 55-60°. My Widowhoods are let out one hour in the morning and one hour at night depending on the weather. If it's raining I don't let them out. I've found that when my birds are right they let me know by not coming in. They muck me about and it's very frustrating. My Widowhood YB food is mixed up the way I want it, I have my food mixed up by Don Blows at Highlex, Strood, Kent. My birds are on breakdown dupurative from Saturday racing 'til Tuesday, then I put them on full mix until Friday, basketing as much as they want. I've found fit pigeons don't eat a lot. I let the Widowhoods bathe Sunday morning. I use Aviform in the water Saturday and Sunday, garlic Monday and Tuesday, multivitamins Wednesday and Thursday, titbits, birdseed and Red Band.

I breed approx 40 YBs. I train these in a few short tosses when they start running, but then on to 30-40 miles Tuesdays, Wednesdays, Thursdays when the weather is OK. YBs are flown to the perch unless they want to pair up, I've found I do well racing to the shed. They seem more consistent raced this way. My YBs are fed just over 1¼ oz a day. I also give the YBs peanuts as titbits. All my YBs are vaccinated with Nobi-vac, treated for canker with Ridzol, treated for coccidiosis and wormed. I do the same for OBs but don't vaccinate OBs as I've found it seems to be a YB disease.

My present lofts are now two Widowhood lofts across the bottom of the garden, 12 cocks one side, 16 cocks the other side. I have a YB shed down the side, making an L-shape, which has a team of 40 plus YB, and a small flying-out stock shed. My tool shed has been converted into a Widowhood hen shed. I'm running out of room and have been eyeing up the dog kennel, I might even be able to fly a Widowhood or two out of it!

One of my best ever pigeons and favourite was a gift pigeon from my father-in-law Bill Wiggins of Canning Town. He was a Silvere Toye, only two years old. He got lost at Berwick and went off course to Swindon. A non-fancier got him in, and had him for two days before finding the wing stamp. I drove to Swindon but was too late, he'd died an hour before I got there. My poor pigeon didn't deserve to die in this way, in a rabbit hutch. He'd won ten 1sts, but his best performances were 3rd Combine, 9th Combine and 13th Combine, up to 9,000 birds. He's now got pride of place on the mantelpiece, stuffed. This pigeon alone won £2,000.

I was given a Delbar cock from a club member, Harry Hughes, and put this with a good Delbar hen from my father-in-law, Bill; this was how I got my Combine winner. To any novice fancier who's reading this, don't overcrowd your loft, keep the loft well ventilated and clean, don't be afraid to try different things. When you're buying birds, go to the most consistent flier and remember, it's quality not quantity that counts.

As a pigeon fancier I do my share for the sport, I'm club secretary, ring secretary, Fed delegate, Combine delegate, chief marker, and a vice-president for the Combine. Coming to the end of my story, I'd like to say I haven't put all my positions down, over the past few years, as it got too monotonous. I will just say that at the beginning my first eight Combines I was in the first 50 eight times. Two of these races were bad and only 100-plus birds on the night. My thanks to Jean Andrews of The RP for the articles on my loft, and to Frank Tasker for a good breed of bird, the Haelterman. I find that the local people don't write about the local fanciers, and nowadays there's no friends in pigeon racing when you're are the top. And last of all thanks to my wife who puts up with me and does a lot for me. I've even trained my son, he cleans out for me, and he's only seven!

LONDON'S DOUBLE COMBINE WINNER

by Terry Robinson of Cheshunt
1st London NR Combine Stonehaven and 1st Thurso,
winner of Tommy Long Trophy Best Average in LNRC races

After receiving a phone call from the Editor about writing an article for Squills I was very pleased, it was something I had always thought I would like to do, but when you actually sit down to start, you realise it's not so easy. I suppose the best and only place to start is the beginning. I have always had an interest in pigeons and looking back my mum put up with quite a lot as we only had a small backyard. I packed up pigeons during my teenage years, and only went back to racing after I had got married and moved to Southbury Road, Enfield in 1970. I started with Heinz Variety and racing in the Hyde. It was through being a novice and having a flyaway one Sunday that I met up with Sid Jones, we struck up a friendship straightaway. As Sid wasn't racing the following year, he generously gave me some old pigeons to break in and race. I did really well with them and it started me off winning.

After a couple of years I started to look around and buy pigeons in. I went to Alf Baker, and bought a daughter of 'The Laird' which bred a good cock called 'The Tryer' which had four Open London NR Combine positions, his best being 6th Open Combine Stonehaven. I also bought pigeons from the London Auctions, one being a blue Dordin cock from Ron Aldridge of Luton which Frank Hall kindly picked out for me. It couldn't have been a better choice as when paired to a hen I purchased

Terry Robinson with the two year old cock who won 5th in the LNRC from Berwick this year.

The main loft of Terry Robinson showing the top section for the Widowhood cocks.

from Alfie Wiseman this pair filled my lofts with winners. They included my good cock 'The Machine' which won fifteen 1sts for me, including 4th Open Combine Berwick over 11,000 pigeons in 1982, he sired my 1992 Berwick winner 'Sparrow Hawk' being 5th Open Combine with over 10,000 birds. Sire of 'The Machine' is also grandsire of my 1992 Thurso Combine winner 'Blue Hawk' and also grandsire of my good pigeon called 'The Combine Cock', because of his five Open positions from Berwick to Thurso. I also had pigeons from Les Massey and Pat Newell which had a hand in the breeding of 'Blue Hawk'.

In 1982 I went with Bob Taylor to visit Arthur Beardsmore. We purchased a pair of pigeons which Arthur said would breed winners straightaway. How right he was, because these must have been the finest pair of breeders in London. They have bred winners racing from 50 to 500 miles, the list is endless. One such pigeon sold to Peter Pedder is sire and grandsire of at least 12 Fed winners, and also 2nd and 6th Open Combine.

In 1984 I moved to Melba in Goffs Oak, which was an upheaval in more ways than one. Les Massey had packed up pigeons and gave me his old loft, in fact, he put it up for me, it was just a question of keeping the birds fed and watered, as I hadn't decided where to put the lofts permanently. It was about this time that John Leisi packed up, and I bought his complete team, including a squeaker, which turned out to be my 1992 Stonehaven Combine winner called 'The Kestrel'.

This pigeon has caused much sorrow as well as joy. In 1988 from Stonehaven, he sat out for 23 minutes, causing Les Massey and myself near heart attacks, but ended up 45th Combine. It's a day I am not going to forget. But he's made up for it, he was even runner-up for the car in

1991, in which he ended up 10th Open Combine Berwick. He is a very versatile pigeon he flies short and long races being 14th Open Combine Thurso (500 miles) in 1990, 13 hours on the wing.

It was in 1987 that I started to get my team of pigeons together. I had by this time relocated my lofts to the bottom of a big garage-cum-shed, but I felt that I was not getting the best out of my pigeons and never seemed to get the right condition on them. I carried on in this way for a couple of years, and then had a good re-think about the loft and pigeons. As I was racing in the strongest club in the Fed, the Cheshunt, and being one of the most westerly fliers, I decided to concentrate solely on Combines, and hopefully have a better chance of scoring. It was then that Les Massey came in again to help. We sat down and started to work out a plan to put the lofts in the roof of the garage; it was quite a big undertaking. We started in the autumn 1990 and it took a month to complete. The most important thing in the loft to me was the ventilation. I wanted a loft with a good clean air environment, and not a stuffy pigeon smell. We did not go for the traditional apex roof, but a flat roof with square chimneys in it, which cost next to nothing to make but has the desired effect.

The lofts contain three compartments 6ft x 8ft, there are 12 boxes in each, but I only use at the most ten boxes, as one contains my nest bowls when I am racing. As I race all Widowhood I find having three separate compartments makes it easier to manage. I try to keep things as simple as possible, as I don't get a lot of time with my birds. I normally start off racing pigeons I want for the Combines at Doncaster, jumping most of my old pigeons straight in without training, but I do train the yearlings and two year olds. I race the pigeons I want for the Combines every race until the first Combine, then I stop them every other week. Perhaps I feed unorthodoxly but I only break my pigeons down on a Saturday or Sunday depending when they race, as I feel too much heavy food after a hard race doesn't do the pigeons any good at all. I feed all my pigeons on the floor, I do not feed in the boxes as it takes up too much time, but I also feel that certain pigeons like different grains and they pick out what they want when you feed on the floor. I buy all my pigeon food from the farm and mix my own; after all that's where the manufacturers buy theirs from.

In 1991 with everything in the loft finished I was looking forward to racing, the birds looked a hundred times better, their condition was marvellous and I had the feeling I could do well. It certainly proved to be a good year, the highlights being 2nd Open Combine Berwick, I topped the East Herts marking station three times (over 1,000 birds), I was runner-up for the car, and took ten Open Combine cards and won eight 1sts in Cheshunt club.

Little did I know 1992 was to prove even better, it has been a wonderful year. Bryan Cowan has always been around watching me time in when I've scored well so he was invited back as my lucky mascot (by the way lads I'm his agent). I scored well in the Berwick Combine, being 5th and I won Stonehaven Combine. With these two good positions behind me, people were saying you must win the Tommy Long Trophy but, as I said to Bryan, you have won no trophy until you time in from Thurso, because to me that's the daddy of them all. I was remembering back in 1982 when I won 4th and 6th Combine, and the same things were being said, and I never timed in. In fact, I did not see a Thurso pigeon for a week. So it was a good feeling when I timed in my Thurso bird, about the time I had predicted, it was only then that I said to Bryan, "I think we are in with a chance of winning the trophy." But once we had got up to the marking station and I found I was in with a chance of winning the Thurso Combine, that really put the icing on the cake. I did win the Tommy Long Trophy, but I feel I must say a word for the runner-up Steve White, he put up a very good performance which would have normally won the trophy, he was 9th, 3rd and 2nd Combine and then won the YB Combine race.

In finishing I would like to think this article gives novices something to think about, especially when purchasing pigeons, never turn a good pigeon down because it is a cross and not a pure bred. If I had done that I wouldn't have had the success that I have had, go by your instinct and not always the pedigree. Plus my two Combine winners are aged six and seven years, many fanciers would not have kept them that long.

Having a look at his own wing 'The Kestrel', the LNRC Thurso National.

THE ODD RING WINNER

by P Kendal

1st National FC Pau

The editor's invitation to write an article for Squills is a great honour for a small time back garden fancier like myself. I am extremely grateful for being given the opportunity to pay my tribute to 'Holloway Boy' and his breeder Derek Parkin.

Winning the NFC Pau Grand National is great but to do it after a holdover in what proved to be a hard race is even greater. Examination of the race result clearly indicates that many of the established long distance lofts with multiple entries found the going much too tough, and this coupled with the knowledge that a number of top fanciers failed to clock in strengthens my belief that 'Holloway Boy's' performance was outstanding.

During a chat in the clubhouse a few years ago Nigel Finch returned one of my rings unused and Derek Parkin said I can ring a squeaker for you. This bird we now know as 'Holloway Boy', a bird of exceptional quality and courage. My young bird team were flying strongly and almost ready to go into the first young bird race when the squeaker joined them, but he quickly illustrated his courage by trying to keep up with them. He missed the earlier young bird races so was lightly raced as a youngster and again as a yearling.

In 1969 with my wife Helen and my father and mother I attended the NFC dinner in Oxford. We sat at the same table as Ken Morey whose pigeon had won the Pau National. My birds had been placed 1st Sect, 246th Open Nantes; 1st Sect, 132nd Open Avranches and 3rd Sect, 47th Open Pau. With great pride I collected my trophies and wondered if I would ever be fortunate enough to own a bird capable of winning the coveted King George V Challenge Trophy sitting on the table so close to me.

It's been a long haul since I started keeping pigeons as a schoolboy and the disasters and disappointments seemed endless at times but easily forgotten in the golden glow of success. My interest in racing pigeons was sparked off when a local fancier removed my fantail's pair of eggs and replaced them with eggs from his racers. The youngsters really looked the part and the seed was sown. Housed in a run with my chickens, ducks and rabbit my next step was to build a loft. Scrap material mostly old doors removed from house renovations were used and this proved to be a very satisfactory school project.

I was given about a dozen birds by local fanciers and joined the local club. Fred (Jock) Hughes the chief clocksetter took me under his wing and got me interested in clocksetting. He frequently picked me up at the nearby crossroads known as the Holloway and he took my birds with his

'Holloway Boy'.

on training. Jock has always helped old and young alike and has gone out of his way to train birds for fanciers even those competing against him.

My father bought me an ancient looking Toulet puncturing clock and my first club race winner was from Rennes, one of the dozen gift birds. My father got the bug and we raced in partnership until he retired in 1981. We joined the NFC in 1967 so my mother said we needed a reliable clock and bought me a new Benzing Printer. We raced young birds from Lamballa that year and christened the clock with 12th Sect, 70th Open. This was a youngster bred from Golden Plover stock birds purchased from Jack White of Wembley.

A two day old gift bird of 'Champion Twilight' x 'Catalonian Prince'

continued at foot of next page

'CLINTPARK LASS'

by Jimmy Graham of Mr & Mrs Graham, Ecclefechan
Winner 1st Scottish National FC Nantes

My wife and I sincerely thank the editor of Squills for his kind invitation
to put on record the performance of our champion pigeon 'Clintpark Lass'.
But before I go into all that I'll give you a wee run down on my experience
with racing pigeons. It all started some 43 years ago when I was only
16. My father started up and was one of the former members of Brydekirk
Homing Club. I got stuck in and helped him to fly and train them but
it was short lived as National Service caught up with me and two years
was spent with the King's Own Scottish Borderers. In this time I did stints
in Hong Kong, Japan, Korea and Northern Ireland. I spent eight months
in Korea with Bill Speakman VC in "B" Coy. Anyhow after my National
Service my father was still racing and straight away I was back into the
pigeons. We flew well in those days winning many club 1sts and National
diplomas with our good grizzle family. I soon got married and up went
a loft in my back garden. My father then decided like my brother John
"Dun" to pack them in and started racing greyhounds.

Since I started on my own many prizes have been won but one of my
best was in 1961 when I timed my good hen SU55S2996 from Barcelona
976 miles. She was the first bird timed into Scotland and had scored five
fine performances before going to Barcelona. I had another crack at
distance racing in 1974 this time sending a beautiful blue cock SU68S528,
a good club and National winner scoring many times. This time the race
point was Palamos 954 miles and he didn't let me down as he won 12th
Sect E, 209th Open from 1,023 birds again being first bird into Scotland
and winning the Gordon Hare Trophy for furthest flying bird in race time.
By the way this good cock was sold to a fancier in Jakarta for a substantial
sum of money.

I waited a few years until 1982 and thought I would have another go
at the distance racing so again I joined the British Barcelona Club and

Continued from previous page

bloodlines fostered by a feeder, with the earlier gift birds and the Golden
Plovers quickly established a winning team. Later purchases from the
Royal Lofts and A H Bennett formed the backbone of the team still racing
today.

Timing clocks are so complex and varied in make that I have found
clocksetting the most interesting of tasks within the club. If the clocksetter
is looking for a novice assistant don't be afraid to volunteer but remember
that clocks require the same quiet, patient and careful handling as do
the birds.

Jimmy Graham.

this time I sent a two year old blue cock SU80S6468. But again it was a new race point, this time being Perpignan 898 miles. There were 712 birds competing and up came 6468 to win 10th Sect 3, 287th Open and for the third time had the first bird into Scotland. That was the only three times I had competed in distance racing.

I am getting away from 'Clintpark Lass' my Scottish National winner, so I'll now give you the run down on the feelings and thoughts that went into my head before and after she won 1st Club, 1st South Sect, 1st Open SNFC Nantes. It all started when she won six club prizes as a young bird and yearling but never a 1st. As a two year old she flew very steadily but never scored. She was sitting ten days when along came the SNFC race from Sartilly (1) and she won 22nd South Sect, 54th Open from 3,310 birds flying just over ten hours, velocity 1352. That's the only race where her velocity was over the 1000 mark. I got her ready for Rennes (3) weeks later and against 4,516 birds she won 42nd Sect, 84th Open with a low velocity of 996. In this race she flew 14½ hours. Last year in 1991 I again entered her at Sartilly (1) with the SNFC and she won 60th South Sect, 102nd Open flying 13½ hours against 3,180 birds, velocity 972. I knew I had a decent pigeon here and got her ready for the last race with the SNFC, this being Sartilly (2) and with 3,673 birds competing and with a very low velocity of 882 she won 27th South Sect, 63rd Open flying 14¾ hours.

I then got it into my head that with four National diplomas and only three years old I would go for a Gold Award (you need five). So she was prepared sitting ten days for the first National of 1992 from Sartilly (1) and I said to myself if she fails I'll catch the last Sartilly. Anyhow my brother Alvar who had bred her came down from Montrose to watch her home. I said to Alvar that it's too easy a day for her, she needs at least 13 to 14 hours fly. The leading birds were homing fast into the south. I timed my Dutchy Cock SU89S4493 at 16.51 to win 8th South Sect, 25th Open. I timed another at 17.13 to win 22nd South Sect, 81st Open with a Decroix hen. I sat a bit longer and had to leave at around 6.30 to take my clock to Annan leaving Alvar with my second clock to time her in. I shouted to Alvar she'll be home just after 7 o'clock and she was. Alvar timed her at 19.27 too late for the prize list. Sitting that night by the fire I said to myself this hen should be at Nantes as she's always there at around 14½ hours. Nantes was only three weeks away and she was feeding a 12-day youngster but she also had claimed the next nest box. After ten days she laid and was sitting seven days and feeding a big youngster that she was very keen on. I never trained her but just sent her with two others.

The story only starts now. It's hard to explain what goes through one's head when you've a good one. Read on. She was liberated at 6am into a light north wind. I thought they'd be an odd one into the south that night as conditions were spot on for some to make it. There was five of us sitting and looking from 7 onwards. I knew it was too early but I also knew she was still flying. Then 14 hours passed and then 15 hours passed. I kept thinking, not saying anything to anybody just looking south. At 16 hours it was getting dark and everyone left to go home. I sat to 10.45 and it was nearly dark. Come on lass keep going I was saying to myself. I controlled my clocks at 23.00 hours and watched TV all night not going to bed. I was shattered as I thought she was good for the day. I thought she had let me down, I thought everything. Of course I didn't know it was a bad race. I didn't phone anyone to find out, I still thought someone had one that night.

Anyhow come 4am I was out looking; it was a lovely morning. Around half past five I saw the village young birds out exercising so I let mine out also. They flew around three-quarters of an hour with the village youngsters then came up to hand. As they were circling my loft I noticed a bird coming right up the field straight from the south. I noticed a gap in one wing and I thought it was one of my youngsters coming away from the village ones. She circled twice and as I was watching her straight down and in. I knew then but didn't know what hen it was as I had two away of the same breeding. Removing the rubbers I noticed she was rung on the left leg. It was her. I put her in her box and she actually played up to me. She was in great condition. I kept walking in and out of the loft for the next half-an-hour saying to myself I knew you'd come, over and over again. I waited a bit longer and then phoned Roy Seaton at Annan

'CLINTPARK LASS'
Blue hen SU88AF00697. Bred by Mr & Mrs A Graham & son of
Montrose. Owned and raced by Mr & Mrs Jimmy Graham of Ecclefechan.
Second Club, 22nd South Sect, 54th Open SNFC Sartilly, vel 1352,
3,310 birds, 443 miles flying 9½ hours 1990; 3rd Club, 42nd South
Sect, 84th Open SNFC Rennes, vel 996, 4,516 birds, 485 miles flying
14½ hours 1990; 5th Club, 60th South Sect, 102nd Open SNFC Sartilly
(1), vel 972, 3,180 birds, 443 miles flying 14¾ hours 1991; 2nd Club,
27th South Sect, 63rd Open SNFC Sartilly (2), vel 882, 3,673 birds,
443 miles flying 14½ hours 1991; 1st Club, 1st South Sect, 1st Open
SNFC Nantes, vel 854, 930 birds, no day birds, timed 06.14 second
morning, won by over two hours 1992. She also flew SNFC Sartilly (1)
three weeks previous, flying over 13½ hours. Norman Orr of Stonehouse
bloodlines.

asking him how many he'd heard of. None was the answer, I've timed
at 6.14am I told Roy. "You've got a good one" said Roy, "it was a hard
race." Then he told me about the English lads and the bad returns for
them. We talked a bit longer than hung-up. I then woke up my wife saying,
"Helen, I've timed her and Roy Seaton says it's the only one he's heard of."
    Well that was the start. Out to the loft and she'd moved to the nest box

where her young bird had been. She was cooing with her cock. I handled her and checked her wings. Her second flight was half-way up on one wing and her other wing was the same on the second flight but her third flight was about half-inch shorter leaving a big gap. How she made it I'll never know. Then I said to myself she'll win the club that will give her her first 1st. I waited to around 9 o'clock thinking and thinking then I phoned again. Jimmy Dalgliesh had one at 8.14. I was told it was only the second bird they knew of.

I walked out the back door and just stared at her sitting in her box. The section was now on my mind as I had two hours on James over in Annan. But not everyone phones in so there could be another one somewhere. Back for an update and they only knew of the three birds into the south. Jimmy Graham of Ecclefechan 06.14, J M Dalgliesh of Annan 08.14, A Simmons 08.48 and this was around 10.30. About one o'clock I gave the phone another go for another up-to-date version and I was still leading by the two hours so I was by now quite sure the Section was mine. I phoned Alvar in Montrose and gave him the news. As the race didn't finish until the third day I had a long wait but I knew within myself I had won the National.

It still hasn't sunk in and as I write these notes she's just hatched two young, the only two I have from her as I gave the two away to friends she was feeding (one of course was switched). I hope I haven't bored you all but you've got to win a National to understand what goes through one's head. Now for her breeding and I'll tell you how she failed to get her Gold Award. My brother John "Dun" saw an advert in one of the pigeon weeklies by Norman Orr advertising late breds from his Gold Award winner and his Section winner. John bought four of them and when John packed them in for greyhounds he gave me the four. I in turn gave a pair to Alvar in Montrose on the understanding that I got the first pair. I did and 'Clintpark Lass' was one of them. Another good one from the pair I kept was SU88S05708. Another blue hen won for me 1st Club, 20th Sect, 99th Open Sartilly as a yearling and 1st Club, 3rd Sect, 13th Open Niort as a two year old. I lost this one at Nantes this year.

Now how she lost her Gold Award and how close she came. Gilmour Bros & Martendale won 58th Sect and £5 Sect prize money, I was 60th Sect timed two minutes behind them, that was how close it was only two minutes. Thanks for taking time to read this. Also thanks to Hawick FC for the nice card.

THE LEGENDARY PRESCOTT BROS

Special Report by J Kimmance, Litherland, Liverpool

Many there have been over the years who have suggested that I should compile a story about the three fancier brothers, the Prescott Bros, whose racing exploits and whose racing machines were to set alight Liverpool and certainly Seaforth HS and the Fed, Championship Club and Combine they flew in. Being that all three are now deceased, with the final removal of all birds in 1979, it meant that it would always be difficult to obtain now as much material and facts needed for a story. I always tended to put it off and still the requests came to do otherwise. This was coupled with an enormous indepth feeling that one day I would, and that I wanted to. Well now I have decided to try, and in doing so, hopefully appease all those who have wished that I would try, but so much will have to come from memory I'm afraid, but what memories, so here goes.

The Prescott Bros were born and bred in Litherland and flew their birds over the years from a couple of locations in Field Lane, but both within 50 yards of one another. So these lofts covered an expanse of very many years and both housed the vast number of winners over those years.

My own very first meeting of the brothers concerned the eldest, who was Tom, and who was the local kid's barber in the backyard in the summer and inside the leanto in the winter, of his house of course. It used to seem like every fortnight from memory that father used to send me packing to get my hair cut, off Tom Prescott. Unfortunately this operation occurred on a Sunday morning, and with Tom you could have any style you wanted, so long as it was short back and sides and plenty off the top, and when I went it was 'two cutting' and 'four bagging'. Indeed from my experiences a little later in life of Army barbers Tom Prescott was a ready made one.

It had to be Sundays for Tom's 'operations', with racing usually on a Saturday, but no Channel birds home on the Saturday, meaning a Sunday wait, then the 'shop' was closed, which for me was a chance to do other things, than the boring, hated sitting in that 'operational' chair at Tom's home. Tom was a very tall big man, always elegantly groomed, and looked the 'perfick' centre half. He would often talk football, naturally pigeons and sometimes his other love, crown green bowling.

With Tom, if he did anything, he did it right, meticulous in everything and with birds and loft especially so, which were housed at his home in those early days. Indeed all was spotless, baskets well varnished, nest bowls and nest boxes, with nothing left to chance and whilst the loft was not palatial, it was dry and clean and sanded every day. So it was not surprising to this writer that they enjoyed the success that they did, for they were to become, in my opinion, better and better.

His brother Fred was a shorter stocky man, 'The Thinker' again in my opinion, and another whose other great love was bowls and a very good local champion at this pastime. Last but not least was John, better locally known as 'John Willie' only because of the many uncles and cousins of the big Prescott family clan of Litherland, this name indicated which John Prescott one was speaking of. John was also a bowls enthusiast, also of repute, but I cannot believe he could match brother Fred ever.

The three brothers and their wives were all very close and all shared a common interest for both bowls and pigeons. Indeed both Fred's and John's wives played bowls for the same Litherland ladies team and were very good. However, there was always someone on hand to do the honours if birds had to go to be race marked or a clock had to be taken for checking. I used to think at those far off times that pigeons should and had to be the only thing in one's life. To me there didn't seem room for anything else, and maybe selfishly, I still think this way, to this day.

Coming back to Tom, and because the loft was sited at his home, he was probably the one who did most for the birds, although his other brothers lived quite close. Also was the fact that all three brothers worked for the same firm which itself was only 150 yards from the loft. So it was convenient for all the brothers to be at the loft, break times, lunch times, evenings after work etc, and sharing the workload in turn, and what a team they were, and so formidable.

As mentioned regarding Tom and his efforts alone, I can well recall that their birds were given toast, crushed up into small lumps about the size of a bean or often to virtual crumbs. Tom would spend time mostly in the winter months 'toasting' and would accumulate masses of it which would keep in this form, until it was needed then crushed. This idea is not a new one and I understand the many lofts 50 years ago fed toast, of course helping fancier's pockets, and especially so during the shortages of the Second World War. Our story actually starts slightly before the start of hostilities in 1939, in fact in 1937, when this threesome won the great North West Combine from Nantes, with a yearling chequer hen. The North West Combine was of course a far greater Combine in those days than it is at the present time, with many Feds, Amals and clubs splintering to other Combines etc, both this side and the other side of the Mersey, certainly over the last two or three decades.

However, this 1937 win was a great win for this yearling Combine winner, and they were to do it again with a yearling which I will try to describe as we now move on through a period of time and the war years of 1939 to 1945. During this period, racing was restricted naturally with races of 200-250 miles to the south coast or Penzance, the maximum distances, dark days indeed.

Come the end of the war in 1945, the race programmes in 1946 were decided on a south east route across England and into Belgium. The birds did not seem equipped for this route or the distances, after many of them had been restricted to just 200 on the South Road, and the result was terrible losses for very many fanciers including Prescott Bros. The year

Prescott Bros (left to right): John, Fred's son Alan, Tom and Fred.

The photograph is 35 years old and all three brothers are pictured as follows: Standing *back row* John 'Willie' Prescott *extreme right*; Tom Prescott (next to Mayor) *third from right. Front row* (with three ladies) Fred Prescott. Also in this picture is a lady who has just celebrated her 83rd birthday, Miss Amy Cowpland, the dark haired lady seated middle, below the lady standing between Tom & John Prescott, in the back row.

1947 was to be the turning point for sheer brilliance for this partnership, and a man called Peter Don.

Peter Don was a fancier who like many more, saw war service interrupt his great love of pigeons, but being a Liverpudlian, he kept his eyes open and was a good listener also. The story goes that Peter Don, towards the latter end of the war, no doubt in early 1945, is alleged to have seen these particular pigeons in a loft in Belgium. This perhaps cancels out the theory that in wartime all the pigeons were confiscated/removed, by the occupying forces, in this case the Germans. Obviously many were, some were not, and Peter Don had heard and found a family, which the Prescotts always referred to them as Costameres, but in all this long time I have never come across in pigeon circles or strain names, any similar reference, none whatsoever!

Whether it was the funny colours of these pigeons that took the eye of Peter Don in 1945, with the blues carrying brown bars, and the chequers, brown chequering, often with the steel grey flights and tails, is something now of conjecture. What is fact, is that eventually over the years, Peter Don made one or more trips to Belgium and certainly in 1947, brought back to the UK and Liverpool, birds of this funny colour.

A friendship arose with our three subjects and Peter Don, a small dapper man, always neatly dressed. I well remember his distinct sandy red waved hair. Peter lived and flew on the east side of Liverpool, part of the great Liverpool Amal and he flew well and was a good fancier of course. Alas he too, is now sadly deceased. Naturally all friendships see an exchange of birds, and one by one, the funny coloured ones began finding new homes in the Prescott loft at Litherland.

Peter Don had proved the abilities of these funny coloured ones, from the time he first introduced them in 1947, and when the breeding really started in the following years. Likewise Prescott Bros, as good fanciers that they were, knew also what they were bringing in. By this time and into late 1949, they had through Peter Don, plenty of pedigree in those that arrived, and I think that was about 1950 or 1951.

I do know, however, that out of their 1956 batch of funny coloured bred youngsters, one was to go on to win another Combine race for them, as I will now try to explain. The Prescott Bros referred to these as mosiacs (chequers) and red barred blues (blues) plus the addition of any white also (pieds). I actually have always thought they were bronze and red barred blues. However, it doesn't matter, they were the goods sure enough, a dual-purpose family that would win from the shortest race, to the longest race, fast or slow races, usually the latter. They were flown Naturally, long before the continuing desires of fanciers of this day and age to race Widowhood.

This partnership used a system that was simple and maybe unique in some people's eyes. It was however different. They believed that in the winter time, and quite rightly so, that this is the time when the following year's races are won. A time when pigeons needed to moult and rest quietly, to be well fed and generally left alone. Although in years to come and as their fame spread, Saturdays and Sundays throughout the winters, saw a steady stream of cars and visitors in Field Lane, so no doubt the birds were disturbed somewhat on these days.

I should add here that in the winter months, or indeed after the end of racing and into September, the birds were parted and they never went out of the loft during the winter months. They were mated the following year, usually in March, with each pair locked in the box until the eggs were laid, thus ensuring parentage. Once the eggs were laid and the pairs more or less settled, the birds would have their freedom of the lofts, inside and out, and in fact, from then on, were on an open hole. In the 1950s the birds were at Tom's home, and he lived in a row of cottages, some ten or so, that were affectionally known as the 'Pig Sty Houses', simply because this row of cottages had their front doors facing the canal bank, with their rear doors actually facing Field Lane, long narrow gardens, and in years before most occupants of these all kept a couple of pigs in their garden, of course hence the name.

However Prescott's loft, whilst covering a long length of the narrow garden, was nearer the gateway onto Field Lane, rather than nearer the house, resulting in the birds being subjected to all sorts of traffic noise and all sorts of people passing the loft, walking down the path to that rear door, which to all intents was their front door and certainly only entry door used.

This traffic served a purpose in that the birds, now on the open hole, were often more in the air than playing in and around the loft, and it convinced the brothers that they used to fly themselves fit.

The old birds received no training in the official sense, but the firm they worked for had a sister company in Warrington, which most days of the week had a wagon into Warrington from Liverpool at least once, if not twice a day. Although off the line of flight, in the accepted sense, with Warrington lying to the east, the brothers always had a basket of birds on each trip, and that's all they received day after day.

The birds received a recognised mixture of corn, nothing fancy plus the 'toast', the odd seed mixture for trapping, and always after the race and Sunday morning, a good helping of Hormoform, which saw the droppings over this weekend period turn a distinct reddish/brown colour.

If, come the first OB race, the brothers thought the birds were not quite right, or there were too many squabs in the nests needing attention or that the weather was too cold, any reason other than near perfection or perfection, would see them miss the first or even the second race, averages meaning nothing to them.

The year 1957 for them started with the same preparation as described and like all other years the inland OB programme saw them winning their share with the funny coloured ones with the sparkling nut brown eyes, and then onto the Channel races.

At Dol (2) they entered amongst their team for that race, a bronze yearling cock of medium size. In the main they were never big pigeons, however, upon basketing at Seaforth LMS station (note rail transportation then) this pigeon was very quickly picked out, as being in superb condition. In fact, the late Dick Woods, also of Seaforth HS was heard to remark, that it would win. Little did anyone know that it would not only win the club and Fed, but also 1st North West Combine, a tremendous achievement fulfilling the highest potential that both Peter Don and Prescott Bros had believed.

But a disappointment for them was to come about, regarding the Combine cock. Prior to his Combine win the bronze or mosaic cock had bred a pair of youngsters. These when raced did nothing at all. He himself never raced again and was heavily bred from. Again his youngsters did little or nothing to shout home about, were easily lost or quickly suppressed. After a couple of frustrating breeding years, certainly as far as the Combine cock was concerned, it was suggested to them that the grandchildren off/from him might be the answer, this it was, with a line of breeding adopted of grandsons mated to grand-daughters, grandmother to grandson, grandsire to grand-daughter, aunt to nephew and vice versa etc, but never cousins nor father to daughter etc, not wanting to take them that close, within the adopted line.

From then on it was win after win, especially over the Channel which for them ended at Nantes. They never liked the 500+ mile races, feeling that it was not a fair race with always very few day birds and winners. However, at one stage they had that many winners within the lofts, resulting in very overcrowded numbers, that it was decided that a cull would have to be the order of that day.

That year the one and only Jim Cookson, mine host at Cookson's pub, who was a clubmate of the brothers, and who also had acquired some of the funny coloured ones, managed to persuade them to send to the longest race Rochefort at 528 miles, and relieve their increasing numbers.

This they did, sending with reluctance ten birds. Jim Cookson was no mug at the pigeon game, had also sent but he could not be present to watch out for arrivals. It was agreed that Fred Prescott would go to the pub, to watch out and time in for Jim, leaving Tom and John at their own loft. You might have guessed the result with more winners for Prescott Bros in a stiffish race, all home by the way, and what about old Fred at Cookson's loft, his brothers had seven home before Fred saw an arrival for Jim Cookson.

They were never afraid to bring in a cross, but always from old established long distance families. Some I remember included: John Cropper of Banks, Fred Law of Marshside, Tom Wilson of Tarleton, great fanciers in their own right, and there must have been others also. I remember also the many names that did well with those funny coloured Costameres including Dick Toman of Moreton, Ringland Bros of Ellesmere Port, Ian May of Bebbington, Austin Flynn (late) of Banks, Crombleholme Bros at Breeze Lane FC, Jim Cookson I have mentioned, and loads of others whose names are now lost in time for me, I'm afraid.

It would be wrong for me, not to mention that through the Prescott funny coloured lines, my own first ever Fed winner occurred, that being also from Rochefort 528 miles, a two year old bronze cock, winning club, Fed and 8th North West Combine in 1960. The Prescotts were 2nd Club and had prominent other positions.

The Prescott Bros funny coloured ones, I am pleased to say are so dominent in our funny colours and pigeons, so that they remain in our loft to this day, and they remain in memory not only of the legendary Prescott Bros, but to the man also, who first introduced them, the great Peter Don. I believe the finest tribute ever bestowed on the Prescott Bros, was one given by Graham Hare of Southport, who was in business in Liverpool, a family concern of locksmiths and security safes. He became a close friend of the brothers, and subsequently by request, was to compile their first ever sale list, which was published in this journal, their last and only other one, being the final dispersal in 1979. Graham stated and I quote him: "That Prescott Bros were champions of the North Liverpool Fed", a statement that has not been disclaimed, nor can it be, because they really were outstanding, and long before many of the present day fanciers. The name of this game, then and now, is to win. Prescott Bros believed this, set out to do so, and did it and very well. They were indeed good for this sport and it is now all the poorer by their absence and they were good for this writer, and it is a pleasure to be able to write about them as a result.

TO DO THE DISTANCE YOU WILL HAVE TO SUFFER

by Des Coulter
1st BBC Perpignan

This may be an unusual way to start an article but I feel this is why probably 90% of fanciers never make long distance fliers. You must suffer the indignity of being pushed down the result sheet in early races, week after week! And if you live in a town you have to see your neighbouring fancier's birds dropping in. Also if they are going training, you feel you must find some to send or get down the road yourself. It is such a fine line between necessary training and burning them out.

In a large town to the east of me there must be hundreds of fanciers flying short or middle distance because they are scared of being beaten by each other, so all the reliable old English strains have gone to be replaced by Continental pigeons. I know there is one good Dutch strain but Holland is as flat as a pancake and there is not 20 to 100-plus sea miles of English Channel or 150 to 300 miles of North Sea as there is out of Lerwick. I must add that distance birds need a fair amount of open loft, so some fanciers are snookered by their loft positions and have to adopt modern ways.

If you can take the stick don't race week after week. Go on holiday; take the family out or just relax in the garden or flat. Your next race will soon come round with a fresher you and a fresher team of pigeons. You may have noticed I am not the best of clubmen but most of the top fanciers in the area fly my strain successfully along with their own and accept me for being as I am.

The two top Widowhood fliers in my local club greatly respect me as I do them, Bob Short predicted I would win the BBC. John McGee has won the Pau National. They both have Naturals as well (fairly Natural anyway). As mentioned in the Pictorial, I feed tic beans plus a mixture of maize, wheat, barley and a conditioner of seeds plus a little cod liver oil once a week. My beans, wheat and barley are bought from local farmers. Sometimes I add an iron vitamin molasses type product to their water. I have never fed a single peanut!

As I expected my bean feeding has aroused a lot of interest. These are fed in hoppers, so the pigeons are never hungry (they trap to a mixture). They must clear up every bean once a day. Straight barley replaces beans the weekend I race or if I encounter a smash training toss.

My distance candidates only get two races before the big event, one inland, one Channel, and are ten to 14 days on eggs. Yearlings have four to five races. They must go to the longest club race — ours is 415 miles or Dax with BICC at 494 miles. Youngsters get most of the programme, but never get trained twice a day.

'No Surprise' GB90N72373, grizzle cock. 1992 — Perpignan, 583 miles, 1st Open, 1st Sect G British Barcelona Club.

My birds do get some training. I train a little to 60 miles but my friend (ex Joe 90) Martin Watson takes them when necessary, anywhere, usually 50 to 60 miles, often out of the London postal area. My Barcelona candidates would have noticed the Welsh mountains on their right when they come out of Stroud, near Cheltenham, their final trainer, thanks to Martin.

I have mixed feelings about the YB National. 'No Surprise' was in the 1990 YB National but was out for months. It is probably better not to put youngsters over the water. The National to me is like a gold rush; thousands go but few find gold worth having.

Eliminate Stress for Long Distance Racing

Whatever age one enters the Fancy one is being encouraged to make your

Left: OB & YB loft. Right: store shed and stock loft. *Photos by Brian Reeves.*

pigeons keener or to go faster. For long distance my observations are to the contrary. How often in a longish difficult race does an unknown name appear, often to the disgust of the week after week sprint boys with their burnt-out pigeons. The unknown pigeons are probably on only their third or fourth race of the season. The hard race to the unstressed pigeon was no problem.

Eggs should only be touched if you are floating them under other pairs or you are giving them to someone or accepting a pair. Clear eggs are unheard of in a well managed loft. Don't pay the sort of money for pigeons that you will be frightened to fly out. Youngsters should not be handled, only for ringing and weaning but should be checked for weaklings and bad feathering. Once weaned, a stressful period in itself, this is probably the only time stress can be useful in the process of the survival of the fittest.

When youngsters start running you know they are bubbling over with energy and looking for adventure, probably flying over new ground every time they exercise. In ancestral times they would be searching for food or future food sources. If carefully trained, youngsters destined for the distance should not be trained in the evening as this worries them in case the don't make their perch by nightfall, causing them stress. They should repay you for this morning-only training with that super performance some fanciers never get with their stressed youngsters or old birds.

If a bit of commonsense is used a team of youngsters can fly most of the young bird programme without coming to much harm. They must be well fed and they will trap from the races if not let up too early. The early morning race liberation in the local federations are losing more youngsters than if they were let go after consultation with other federations at around midday from the first few YB races.

If the races are hard, don't expect too much of them for a few days. I can remember one fancier, whose youngsters had flown 170 miles on the Saturday, flying 6-7-8 hours, was surprised when he lost some of them from 45 miles on the following Monday! A night out won't worry the unstressed long distance youngster. Next week it may even win.

Never send to the race on the Saturday before basketing for an old bird or young bird Classic. Most comeback races are unnecessary as are 350-mile races. Leave the comeback races to the Widowhood boys. A trainer around 40 miles or a couple of 15 to 20 miles is OK weather permitting.

Yearlings must be entered in one or two of the early races, 70 to 130 miles. You don't have to breed from them. They are after all an untried section of your racing team, no matter how well they are bred. With a bit of rest and a 60-miler they are ready for the final Channel race. It does not really matter how well they do, remember, I am looking for an unstressed pigeon that is going to win that super position in a Classic or an International club later on.

Two year olds and older if they have been treated as I have advised and then you should be laughing providing you have long distance stock preferably of British descent. This could be getting hard to come by! Then with occasional careful training, as much open loft and food as possible and the magic formula (this is not a potion) you could score in a number

Wing of Perpignan winner. When bird arrived it had seven flights to go.

continued at foot of next page

MY LOVE FOR PIGEONS

by Les Nicholls, Bristol
1st Open Nantes NFC
As told to Bristol Bulldog

I would like to thank the Editor for asking me to tell the story of my pigeons and the way I keep and fly them. I know this comes about because of my winning a National race (and a car) and I can say I am proud to be ranked alongside the many famous fanciers who have had this honour in the years gone by. I will start by thanking friends Roger Keepin and Alan Chaffey for help in compiling this article.

Born the last of ten children into a family of livestock keepers and dealers I have always been in contact with animals and birds. My grandfather was a blacksmith with a smithy at the end of the garden of the house. My father was what was known as a "general dealer", and it would be right to say he was a "character", very well known in the Bristol and surrounding area. As a youngster I had all sorts of livestock with rabbits, chickens, goats and ponies prominent and the goats were my favourites at this period.

One of my uncles kept pigeons and gave me a few pairs (definitely of mixed variety) to keep. I was now hooked on pigeons and eventually after some years decided to "do it right", so bought a 21ft x 6ft loft made of packing case wood for £21. Now started the quest for some proper racers and a local fancier was selling up so I called on him. He certainly had some great looking birds and I bought a lot from him. This was in 1972 I think, and they bred very well in my large loft. When I started them training etc, the losses started immediately and kept on right through my first year with them.

I know now why this was, for the "good lookers" I had bought were basically show racers, and of course "went down like flies" when put on the road. This experience put me off blue coloured pigeons for many years, and although "I still like a good looker" I like to think I also know enough to find a "racer" also. I replaced these early disasters with birds purchased at sales at the old Birmingham Mail Show, local auctions and from local fanciers John Glover and Alf Burgess, both now sadly no longer with us. These birds now set me on the right road as a racing fancier.

Continued from previous page

of races which I will let you work out for yourselves.

Hoping what I have written may help someone to join the queue at Salisbury and have a 1st out of Palamos one day. I must thank the Editor for asking me for an article for Squills, also my good friend Jed Jackson for theory back-up and Martin Watson for his invaluable help training and my good wife Pam, who looks after the birds when I am away.

As Jan Evert makes his short speech the rain spots start, but nothing can dampen the smile of car winner Les Nicholls.

One thing my birds had to get used to was their changing surroundings. The garden at home was something like 120ft long and father had developed a tyre business. The garden was a tyre store and completely covered the garden at times ten tyres high right up to the loft. The pigeons had this to look at all week when coming from training etc. Come Friday night the birds went to the races, and the tyres were all put on to lorries to go to markets on Saturday. Home came the pigeons to a completely different looking garden, not a tyre in sight, but they trapped when called and cards were starting to be won in the Bristol West FC on the North Road. When I hear of pigeons being put off by a chair left in a garden etc, I think of the comings and goings of all those tyres and smile to myself.

This was the way I flew pigeons for some five years, with father making the old helpful comment like: "If they was chickens at least you could make a meal out of them". I had built up a large team in this loft with the young birds and old birds all kept together. This sometimes meant that the birds left their pens and nested on the floor, leaving the youngsters in the pens. By now I was also following Nature's trail of courting Lyn

'Blue Fiesta', 1st Open Nantes NFC 1992.

who then became my wife and we started our family. Another new move was to live above and run a shop that father had taken over.

The pigeons were later moved to my allotment about ½-mile from the shop at the top of a hill. I could see the birds at exercise from the shop. One hen, a daughter of 'Banana Queen' was so tame she would come to me and pitch by me if I was outside the shop whilst she was flying out. This hen really won a lot of cards including four 1sts. I will now explain the reason why my best hen at this time was called 'Banana Queen'. I had some birds from old fancier Alf Burgess who had been very succesful in his day. The hen of this pair did not lay for a long time, but at least when she did lay one egg it was a large one and badly misshaped, like a banana. My mate Jim Hartrey laughed when he saw it, saying it would never hatch.

I had already satisfied myself that this funny egg was fertile so I just said if it hatched the youngster would be called 'Banana' — and so the hen was now 'Banana Queen'. She won as a young bird and as a yearling produced four 1st prize winners so was never raced again. I had a few bantams at this time and put some eggs under pigeons to keep them from laying all the time. These eggs hatched and where the pigeons usually just left the chickens, 'Banana Queen' continued with the chick, following it all over the floor of the loft and sitting it at night. This is where I got the idea to try different things with sitting pigeons. Her children were always very tame regardless of the cock paired to and I have had birds bred off her come from races, hit the loft and fly to me, then I had to go into the loft to put their rubber in my clock.

The success I won soon brought the attention of thieves and my lofts on the allotment were broken into several times with pigeons and equipment taken. I got really disheartened by these constant break-ins so asked Alan Chaffey if he would sell them all by auction. I was so down in the dumps I advertised an entire clearance sale and all went. Within two years Lyn, myself and our daughters moved out of the shop to a house with a garden, and soon the chickens and ponies I had gone over to, began to be replaced in my mind with pigeons. Soon a loft appeared in my garden and pigeons were re-introduced. Some of my original birds were obtained and others came from local successful fanciers. Unable to race for another year, owing to my sale I used this time training my pigeons and occasionally had a clock set to help compare my times with the other members.

So a team was partly established and I resumed racing in the South Bristol IFC on the North route. Eventually I joined the Bristol West FC

Les Nicholls, left, holding 'Blue Fiesta', 1st Open NFC Nantes.

Les Nicholls with double grandson of 'Rainstorm'. Les bred and raced this cock with great success and now at stock he breeds winners every year. 1992 sire of 1st St Malo.

as well so competed in two clubs of over 50 members. I really like to win cards, whether racing or showing and so was delighted when the birds started getting into the frame in both clubs. The birds I had at this time were really honest pigeons and I still have some of this blood in my loft today. Eventually though I felt need of change and decided to get some Verheyes.

I paid many visits to LPW and bought what I thought was the best on offer. Now being a lorry driver in the family haulage business and having a family I bought to the limit of my pocket (and more) but only if I felt the bird would strengthen my loft. I left one of the NR clubs in 1987 to try my hand on the SR and also joined the NFC that year. My birds flew well enough in the Hartcliffe & Dist SR Club of 60+ members and my first attempt at NFC was that year's Nantes race when a blue yearling Verheye hen was my only entry. She won 10th Sect G, 23rd Open, and she was the sister to the grand-dam of 'Blue Fiesta', 1st Sect G, 1st Open Nantes 1992. This 10th Sect winner was christened 'Lamb Chop' because when she came I had just started to eat my dinner in the garden. I jumped up, went to the loft to time her in and when I came back my dinner plate was empty and my dog had a look on her face to show she had enjoyed my dinner of lamb chops and veg.

I also won the YB Av in the Hartcliffe & DSR that year, but will not bore anyone with lists of wins, although I have every card my pigeons have won. I will just say I have won 1st prizes from every racepoint of the West of England NR Combine and the West of England SR Combine over the years and have the cards here to prove it. I tried one season of Widowhood flying and won nine 1sts but to be truthful I could not stand the sight of all these well bred hens doing nothing. When dad heard I was on the Widowhood method he offered another of his gems: "I have been on Widowhood for eight years and it ain't done much for me". This was a final comment and I abandoned the Widowhood style there and then, and returned to the normal ways of pigeon keeping.

Now to the pigeon that is the reason of this article, 'Blue Fiesta', 1st Sect G, 1st Open Nantes NFC 1992 pooled in everything to win a new Fiesta car and £2,283. I am told I have a way with livestock and try many things with pigeons. 'Blue Fiesta' and her mate left their pen after the first nest and built their second in a corner behind a long handled scraper and brush. When sitting she would not voluntarily leave her nest, the cock having to force her off to do his stint of sitting. If I lifted her off she complained vigorously and when put on the floor ran as fast as she could to the nest. I wondered if she might be even keener if she was sitting something out of the ordinary, so replaced her eggs with a chicken egg. They sat it and 'Blue Fiesta' was very keen indeed and so it was left this way.

My pigeons are trained three times a week from "The Gate", a field

Double-decker 'L'-shaped loft of Les Nicholls.

entrance gate at Brent Knoll in Somerset, about 18 miles away, and this hen was trained the same way as well as racing to Redon in France with the club. Some Multivits are given, Johnsons Tonic is used and I buy Cow Minerals (for use with milking cattle) like pink minerals only cheaper, and a mate gets me stone dust from a Mendip quarry to use for grit.

I feed a mixture of peas, maize and red dari for racing and breeding and there is always corn in my feeders, a thing that seems to amaze visiting fanciers. During the close season they go on to 60% wheat and normally enjoy good health and moult. I do worm them prior to pairing and that is about it. I am not a fervent loft cleaner, for my lofts are dry and pigeons should not be a chore, only a delight.

These days I fly with the Broad Plain House SR and the Hartcliffe & DSR, two top clubs in the WESR Combine and I would like to thank Kay Blake, secretary of Broad Plain House SR, for organising a coach to my car presentation at the Ponderosa Stud and to the members who came along, and also of course to the Ponderosa Stud and NFC for giving me the opportunity to win the ultimate prize — a new car.

Maybe my wife Lyn, who has spent quite a bit of our courting time at other peoples' pigeon lofts will now find it was worth while, for the car is hers and with five young daughters to look after I find there is no room for me in it. Sorry if I have rambled on too long for I can truly say I did not expect ever to do this sort of thing, but could not refuse the honour of being asked.

J&D LOFTS GOLD CUP WINNERS

Jez Hoult & Dave Benison of Rotherham & Brinsworth FC
by Dave Swales

It gave me the greatest of pleasures when I asked Jez Hoult & Dave Benison, the J&D Lofts partnership, if they'd contribute their story of success for Squills. They invited me to their loft set-up for a look round and of course to see and handle the pigeons that have taken Sheffield & District Fed by storm over the last five years. These two young men are more than worthy of any praise I can give them, not only have they been premier prize-winners of their flying club, Rotherham & Brinsworth, 20 members strong, averaging 200-300 birds a week, along the way they also won Federation Averages, Inland, OBs and YBs. But their main achievements for me, that stand above any others, is that they've won the coveted Sheffield Gold Cup for Federation Channel Averages for the third time in four years, also being runners-up the year they missed out. These achievements must be on par with any other outstanding performances that I've ever heard of within Europe, particularly that we fly into one of the most awkward parts of the country, either classed as North Midlands or South Yorks & Derbyshire areas on being so close to the Pennines and having to cross umpteen other flying routes!

Their story begins right from their nursery days, for they were brought up together in the same neighbourhood of Brinsworth, on Ellis and Duncan streets, only a stone's throw from the allotments where the J&D Lofts are now situated. Jez has always been in and around the allotments and the local pigeon lofts and fanciers, as in his infancy and youth he helped his father and elder brother Ray Hoult. Ray still flies at the old lofts, right next door to J&D Lofts. Dave Benison must have been smitten by the bug in the early 1980s and in 1986 the two of them decided to team up.

They started out much the same as many, many more in the Fancy with gift birds from locals and friends also going to sales etc. They would like to thank Ray (Razo) Britland, Ray Smith, Keith Riley, Jez's brother Ray Hoult, Joe Bilton, for gift birds when they started, Birds were purchased from Walt Greaves, the 1986 Gold Cup winner and a few from Louella. All went in by the trial & error method of what is now a very much treasured stock loft! Some of their main attributes are still there today, for example, when Joe Bilton had to call it a day through ill health, he held a sale at Brinsworth WMC in 1989 where 32 lots were sold. Joe had quite a reputation over the years within the area, when the Sheffield Fed carried a lot more members than it does today. The main breed of birds at the sale were Vanhees. Jez and Dave purchased three or four at the sale which were Joe's original stock birds purchased from Louella years previously.

With these was a gift bird from Joe to Jez in 1984, a blue pied hen GB84T88022, a Vanhee. She wasn't raced as a young bird, but later won 1st Club, 3rd Fed Fareham; 1st Club, 1st Fed Niort; 2nd Club, 15th Fed Rennes; 3rd Club Fareham. She was barren as a yearling, but as a three year old she single reared a good pied hen, which become invaluable at stock.

Their Widowhood team has been built up since that time by purchasing some Louella Verheyes. Jez told me that they bought a young bird out of 'Pinocchio' paired to 'Ashmead', also one out of Champion 'Classic Action' paired to 'Zoom Away'. These young birds became multiple prize-winners and in 1988 they went back for some the same way bred! On arrival at the stud they were informed that 'Pinocchio' and 'Ashmead' were for sale and without hesitation they were bought because for the last five years they'd bred, in the single flight set up, winners and producers of winners for numerous fanciers.

At this same time they purchased a direct son of Champion 'Motta', a Vanhee which had bred winners at 500 miles! And their trial & error pairings with the Verheye x Vanhees, resulted in their current main breeding stock birds, chiefly responsible for winning the Gold Cup in 1989, 1990, 1992, runners-up 1991. Some notable names for the Verheye enthusiasts from their pedigrees are: Champion 'Rainstorm', 'Mr Rainstorm', Champion 'Speedway', 'King Billy', 'Action Man', 'Barbarella', 'Black Toney', 'Rainwarrior', 'Rainbeau', plus many more notable pigeons that have produced the goods over the years.

J&D race Widowhood cocks and they all go to every land race, also, most of them to every Channel race with at least two 30-mile training tosses a week, and for the big events, they're taken 80 miles and singled up! They are all raced through from 50 to 500 miles-plus. Their feeding method is the conventional Widowhood way of: depurative, Saturday on return from races; Sunday, Monday, Tuesday, starting to build up; Wednesday right up to basketing on Friday night, as much as they want! As the distance of the races increase, the build-up is brought forward to coincide.

The young birds, 20 cocks, 20 hens, are flown separated for the first three races, then let run together for half-hour before basketing. After the first three races they are encouraged to pair up, and ten to a dozen nest boxes are opened up for them. It works as they topped the YB Classic from Picauville doing it this way. In their own words — they "hammer them all the time!".

Now for some of their results and notable performances and achievements. They first raced together in 1987 — racing ten OBs, winning one 2nd BS nomination in the club. With YBs they won the averages with two 1sts, two 2nds, two 4ths. In their second year together, 1988, winning Central Marked Avs, YB Avs, along with five 1sts, three 2nds, five 3rds, one 4th. 1989, winning Central Marked Avs, Combined Avs, YB Avs, plus Sheffield Gold Cup along with five 1sts, three 2nds, three 3rds, two

4ths. 1990, winning OB Inland Avs, Central Marked Avs, Combined Avs, Sheffield Gold Cup, along with three 1sts, four 2nds, two 3rds, four 4ths, only bird on day of liberation in all Federations from 500 miles.

1991, OB Avs, OB Inland Avs, Combined Avs, Central Marked Avs, runners-up Sheffield Gold Cup, along with ten 1sts, three 2nds, three 3rds, six 4ths; 1st, 3rd, 12th, 21st, 30th Fed Royan, 537 miles; 1st, 10th, 18th Fed Rennes, premier prize-winners inland at Fed level. 1992, winning Continental Avs, Combined Avs, Sheffield Gold Cup, along with seven 1sts, four 2nds, three 3rds, five 4ths, along wih 4th, 6th, 11th, 18th Royan, 537 miles, only 23 in race time; 2nd, 24th Fed Rennes; 1st Fed Picauville YBs.

On handling the racers and stock birds which are small to medium, I noticed that at least 80 to 90% of them had the violet or pearly eye, was it a coincidence? As the locals know, I'm a "nutter" for eyesign etc. The answer from them was that all their top performers had these colour eyes. It's in the family they've created over the past few years. Dave took me into the stock loft to see and handle their main breeders.

To get prolific Channel birds, they must be sent across either as YBs or at the latest as yearlings so they know right from the start of their racing careers what they have to endure and what the Channel looks like. They've lost several good land birds, sending them over too late in life. Their method is teaching them the game young, so they don't forget it — and doesn't come as a shock when put to the task. Teach them these habits and they'll never forget them. "Don't try to teach an old dog new tricks" is the saying.

They said that the Dave Allen Widowhood Book was a big help to them when they decided to race as they now do! Titbits come in the way of Red Band — and every night the old faithful pinch of linseed — they use no "bionic" potions or additives in the water or on the corn. They never ever take anything for granted with the welfare of their pigeons whatsoever. You only get out what's put in!

Once again to our area's "top guns" — thanks for contributing to Squills and a very big well done from all in and "Around the Brook". They are a credit to our sport.

PATIENCE

by George Dobb, Sutton in Ashfield
1st NRCC Lerwick

When asked by the editor of The Racing Pigeon if I would contribute an article for Squills 1993 as a result of winning the King's Cup from Lerwick, I told him that six years ago an article appeared in the Pictorial by "Ted" on my pigeon career and that what I did today was not much different from my methods of that time. However, he stated that the fanciers coming into the sport today would not know about that and many were mature people, who had either retired or become redundant, and that the days of the young lad entering pigeons were few and far between. This is undoubtedly so because there is so much ready-made entertainment available these days. In addition to which the disappearance of terrace houses and semis, with their long back gardens, and their substitution by ugly, inconvenient tower blocks has proved to be a gargantuan mistake by local authorities.

How can young lads be expected to follow healthy, outdoor pursuits like pigeon racing etc, if they are forced to live in the architectural monstrosities which pass for housing accommodation today? There is no wonder they are bored out of their wits half the time, and this is undoubtedly the reason they try to escape by resorting to glue sniffing, drugs etc. The creation of a dependency state by all governments since 1946 has unwittingly led to a feckless attitude towards life. I could go on for ever about the evils of collectivism — look what's happening in Russia and its satellites for instance — but I'm supposed to be writing about pigeon flying so I'll change tack.

The first requisite of a potential successful pigeon fancier is to be blessed with an innate desire to have pigeons around you. If you are in any way lacking this feeling there are so many pitfalls and unexpected mishaps awaiting that only those with the stoutest of hearts should venture into the arena. If, however, you have the ability to overcome the potential misfortunes that abound and which you are, by Murphy's Law sooner or later bound to encounter some, go ahead and, depending on your age, you can expect to have contentment and many years of temporary relief from the worries of the outside world which, rest assured is worth a king's ransom. I started to keep pigeons at the age of five when I first went to school, and even managed to keep them during the war as I was in the RAF Pigeon Service, being in charge of lofts in Yorkshire (2), Iceland and Scotland, but alas the eight birds I left to join up were reduced to one before I returned.

Having now got over the introduction, which can be more fully appreciated by referring to "Ted's" article in RP Pictorial, No 219, Vol 19, I will endeavour to try to enlighten the imaginary newcomer who, I now take it, is still with us having passed the stringent qualifications for pigeoneering to which I have referred. It is assumed that the mature newcomer will already have obtained suitable premises in which to house his stock, and will in all probability have decided on the extent of flying which he intends to indulge. There are, in all sports, horses for courses, but in this respect I would point out that although long distance pigeons can be persuaded to win at shorter distances, the reverse is more than likely to prove disastrous. It behoves one therefore, unless it has been decided to limit oneself to races of up to about 200 miles, to select a family which will go any distance, and to obtain the services of a reliable and successful fancier to help. If youngsters are purchased there is a good chance that at least 75% of them will have disappeared by the end of the young bird season, so that apart from some painful experience having been acquired there is not a lot to show for the time and money expended. Consequently, if it can be afforded, I feel it is preferable to purchase initially youngish, tried and tested stock, or young from parents known to be producers of winners. To avoid unnecessary disappointment they should be kept in a loft with an aviary, preferably with a grill floor which can be cleaned from outside with an instrument like a good garden hoe. Ideally the grill should be constructed in 2" squares and fixed so that predators like rats etc, cannot gain access. This will no doubt tax the imagination of the mature starter but is quite possible to achieve. Perhaps a permanent pest repeller would be invaluable.

I now come to what, in my opinion, is the most fascinating side of the hobby and that is breeding. It has always absolutely amazed me that the vast majority of successful fanciers, some credited with being the best in the country and some self-styled, when asked about their breeding methods state quite dogmatically that they always mate the best to the best as being the one and only road to success. In my experience, unless the best breeders have already made themselves known, nothing can be either more misleading or further from the truth. My King's Cup winner was bred from two pigeons which have never seen the inside of a basket, in fact, the dam was ten years old when she bred him and has never flown out. The sire was a late bred two year old, the first time mated, whose elder brother had won well, unpaired, as a youngster, being 2nd Northallerton; 1st St Boswells 204 miles; and 1st Sect, 2nd Open NRCC Berwick, winning £440 and a clock, on a day when he decided to clear off before entering the loft as I had nothing out to encourage him in.

This cock is actually the sire of my winner from Ripon and was bred from an aunt to nephew mating, which are mated again this year having been separated for several years. Both of them have bred winners with several other mates and can therefore be regarded as a tried and tested pair of pigeons. They are of the family of Busschaert, being very close to Alf Wright's 'Studtopper' and 'Smokie', on both sides. This pair are now at LPW, indeed I had the very first one they bred at LPW and still have it, with the addition of a sister, and the blood of 'Shy Lass' and 'No 1 Cock' via my friend the late Danny Challis. Very many winners at all distances have been produced by keeping this blood intact. The vast majority of my winners and producers of winners have emanated from related parents; indeed although I have had winners as a result of outcrossing, I have not been particularly successful with outbred stock over a period, either for racing or breeding, but particularly with the latter. I understand that most champions are the product of outcrossing and can only conclude that this is because the vast majority of pigeons are produced by this method in the mistaken belief that consanguinity leads to inevitable weakness and disaster. I have no doubt that this could be so in the wrong hands, but practised correctly and assiduously culling the risk becomes less and less and the stock improves overall.

I have read many, many articles over the years extolling the virtues of outright crossing and the Belgians are invariably held up as the masters of the art. From what I can gather they all race from the south, and with the prevailing SW wind and flat terrain over which they fly it is not difficult to understand why so many birds of immediate Belgian origin fail so ignominiously when confronted with 250 miles and the vagaries of the weather and mountainous terrain of the British North Road. I have read that the route must be bred into them and this is no doubt the reason why the Busschaerts, which are comparatively unknown in their country of origin, have made such an impact over here. After 25 or more years the environment has been bred into them and they have been improved and made to respond by the efforts of British fanciers. What they have achieved is beyond comparison with any other British family and I feel they will be around for a very long time yet. My own family of Busschaerts, which are really my breed now, has been cultivated from champions produced and raced in this country — 'Shy Lass' (Tom Larkins), 'Studtopper', 'Smokie', 'Clapper', 'Coppi Cock' (Geo Corbett), 'No 1 Cock' and 'No 2 Cock' of Danny Challis, which are sires of 'Moneypacker', 'Kitten', 'Blue Star', 'Consistent', respectively and many more champions have all figured in the current breeding. I have one or two others but as yet they have not proved to be of any great significance and are therefore for the present discounted.

For further information on breeding I would refer the reader to the resurrected article by the late Dr G H T Stovin in last year's Squills. Dr Stovin was not a winning pigeon man as such, but I have his original book from which I have gained much inspiration. Having now handed over the breeding of racing pigeons to Dr Stovin the next factor in importance in the successful management of a loft is the feeding. I dealt with this in the aforementioned Pictorial, and it was also dealt with by a contributor shortly afterwards. He did not, however, mention where he had got his information from, but it was obviously culled from the late J Kilpatrick's book "The Thoroughbred Racing Pigeon", of which I still have a copy. In it John Kilpatrick sets out the way to calculate the percentage of protein in a mixture, concluding that 16% of protein in any mixture for racing was the minimum requirement for optimum condition, or words to that effect. From my experience I would suggest 15% is adequate, a percentage point or so either way being seemingly insignificant, but by the look of some continental mixtures about 12% of it would seem to suffice.

I know of fanciers racing their young birds on what is known as Depurative, and from the handling of their birds at marking I can quite believe it. They feel like a bundle of feathers, and from their performance in hard races act like they are. However, I still believe that a lot of money is wasted on feeding pigeons, and as for all the supplements and medicaments recommended there aren't the number of days in a week to accommodate them all. I recently read a booklet brought along by a fancier for me to look at. After recommending all the antibiotics and supplements supposed to be absolutely essential, the writer implored the reader to make his birds fly, whether they wanted to or not, for an hour morning and evening. This, despite all else, is really the crux of the whole matter, but I have proved conclusively that 40 minutes twice daily is quite sufficient. They will do this fast and revel in it, and that is where the benefit comes from, not just flopping around the chimney pots.

I next come to the training of tyros (young birds). I always ensure that they have been flying well at home for a couple of months or so, clearing off and getting mixed up with other batches etc, and flying for anything from 40 minutes to some hours. Having reached this stage it is a waste of time taking them any less than five miles as they know where they are and go off for a fly before thinking of returning home. If you want them home quicker let them have their fly before taking them. From five they can go to ten miles, preferably for a few times until they head for home at once, and then 20 a time or two. From 20 they go to about 30 or 40 and then mine go to Leeds which is 48 miles. From there they will fly Berwick, 184 miles, if they're any good, but normally they have about three or four intermediate races before getting that distance.

Regarding training for old birds, for years I have started them at 20 miles as for several weeks before commencing training they have flown well around home and are fit enough to fly at least 200 miles. I know of fanciers starting old birds at two miles and proceeding in minute stages to get them fit. It reminds me of golfers starting the year with a few holes and gradually working up to more for fear of some calamity befalling them if they proceed too quickly. This is because they have sat about all winter and in these circumstances I would recommend they be given 40 minutes twice a day and fed on barley porridge!

For those enthusiasts who are most anxious to do well I would recommend getting hold of such books as "Secrets of Long Distance Pigeon Racing" by Squills, and J Kilpatrick's book aforementioned. I know they are old and possibly out of print but libraries might help, or perhaps older fancier friends might have a copy. These books are invaluable reading and readily assimilable. They are well worth going to some trouble to obtain. *(Kilpatrick's book is still available at £6.50*

continued at foot of next page

RON LAWS OF CHATTERIS
ALWAYS HAS LERWICK IN MIND

by Tithe Rambler

A fancier who keeps and races pigeons, especially for Lerwick, is my near neighbour Ron Laws of Chatteris, Cambs, and it is significant that he won from Lerwick the year I started to race pigeons in 1951, and has again won from there in 1992, over 535 miles. Let's start at the outset, by saying, when I first became interested in our feathered friends, 41 years ago, I visited a number of local fanciers to see their lofts and birds, and to find out as much as I could on the sport.

At that time "Wenny Estate" situated on the south-east of Chatteris, had several fanciers, including Ron Laws, competing from there, and although he was a young man at that time, he had an established loft, and was in fact winning from the club's longest race point, Lerwick. Moreover at that time Ron was indeed fortunate to have met a good Peterborough fancier, E Stevenson, who let him have some of his best Gits, a strain that would not only win from the distance, but shone in both easy and hard races. At that time Chatteris DHS competed in the old Northants Fed which comprised mainly of Cambridgeshire clubs, and did so until it was disbanded following the 1963 season, with the club then joining the strong Peterborough & District Fed. In 1951 Ron won 1st Club, 4th Fed Lerwick, 534 miles, and the same blue chequer hen won 1st Club, 1st Northants Fed in 1952, a very hard race. Then in 1953 she was entered in the NRCC Kings Cup Classic from Lerwick, and again she obliged with 12th Open, another very tough race, won by H Harding of Low Fulney, near Spalding, vel 793. Ron's younger brother David also raced pigeons at that time, having a small loft next to Ron's, and in 1953 he won 1st Club, 1st Northants Fed Lewick, yet another very hard race with few birds home in race time, and his winner was a dark w/f cock also of Gits bloodlines.

Following that successful season, the brothers joined forces, and raced

Continued from previous page

and Secrets was reprinted in Squills Year Book 1988.—Ed.)

To explain the heading of this article, the best advice any aspiring pigeon fancier could possibly be given is to have patience. Most losses of previously good birds have been caused by lack of this precious commodity, and anyone doubtful of his ability to exercise it should paint the word in letters a foot high on a board and nail it up in a prominent position in the loft.

I wish all readers a happy and successful 1993, and would like to thank my son George, who is not pigeon minded, for the occasional training toss he has given the birds. There are no holdovers when he takes them!

Ron Laws seen holding his 1992 Lerwick winner 'Little Gem 2' and a fine array of silverware won.

together for a few years, and in 1954 David's Lerwick Fed topper was entered in the NRCC Lerwick Kings Cup Classic, and much to their delight, duly obliged by winning 47th Open. Moreover, that year saw the partnership emerge as the runners-up in the Northants Fed prize list, being beaten only by the renowned Cooper Bros of Cambridge, who owned one of the best lofts in the area at that time, although only a small one it was one of quality.

After a while, Ron, who was now married, moved to the other side of the town, where he still lives, keeping pigeons for a time, then both he and his brother David called it a day to concentrate on their agricultural "Gangmasters" business. After a few years Ron decided that he would like to race pigeon again, and a loft was erected on his brother-in-law's land (as he had little room to spare at that time) not far from his house, and fresh bloodlines were acquired, from such as the late Wilf Richards of Histon, with others from Mr & Mrs Fred Joyce of Chatteris, with a visit being made to the dispersal sale of T L Pick of Bourne, Lincs, and more recently there has been additions from LPW and the Old Comrades shows, and others. During the past few seasons Ron has done very well for hmself, clocking winners from the shortest to the longest race points, including Bubwith, York, Berwick, Perth, New Pitsligo, Thurso and Lerwick. Without doubt the 1986 season will go down as one of his best ever, mainly because of his achievements from the Shetlands winning a large amount of cash, also a number of diplomas, trophies and other awards. The birds mainly responsible for Ron's success were 'Little Gem' and 'Charlie's Boy', his first two arrivals from Lerwick, flying over 535

'LITTLE GEM 2'

Four year old blue hen, bred and raced by Ron Laws of Chatteris, of Delbar-S B Cooper bloodlines, in 1992 won 1st Chatteris DHS, 1st SE Sect, and 11th Peterborough & District Fed, with 1,259 birds, also 1st Sect F and 121st Open NRCC Lerwick, 535 miles, vel 1104, with 4,450 birds, winning £230 cash.

miles.

'Little Gem' was a blue hen bred in 1982 of Warrington x Delbar x S B Cooper blood, and as a YB she was lightly raced, then as a yearling flown to Perth, 307 miles, and as a two year old she flew Thurso, 447 miles. In 1985 she was given two short races, then into Lerwick, to win 2nd Club, 5th SE Sect of Fed, also 14th Sect F and 138th Open NRCC Lerwick Kings Cup Classic, vel 1177, with 2,773 birds. Prior to Lerwick in 1986 she was again given two short races, then on the second morning of the Lerwick race, his good friend and helper Charlie Coulson timed her in at 6.40am in very misty conditions, to win 1st Chatteris DHS, 1st Cambs CC, 2nd SE Sect, and 10th Open Peterborough & Dist Fed, 1,331 birds, also 2nd Sect F and 30th Open NRCC, vel 841, 2,861 birds, winning also a quartz clock, Harkers Special Tankard, 2B Nom and £905 cash.

'Charlie's Boy' was a blue cock bred in 1983, similarly bred to 'Little

Gem' and was in fact paired to her. As a YB he too was lightly raced and as a yearling was raced to Perth, then as a two year old to Thurso, winning on the way 8th Sect, 75th Open NRCC Perth, vel 966, 2,565 birds, also 6th Club Banff, 374 miles. In 1986 he was given four races including NRCC Perth, then from Lerwick was timed in at 9.55am by Ron's wife Joan to win 4th Club, 4th Cambs CC, 8th SE Sect, 60th Fed, 1,331 birds, also 9th Sect F and 121st Open NRCC, vel 710. Both birds again flew Lerwick in 1987 with 'Little Gem' 7th Club. In the three longest races, Ron won 2nd & 3rd Fraserburgh, 372 miles; 7th Lerwick; and 4th, 5th, 8th & 9th Thurso. It is significant that those four birds from Thurso arrived home together, after 10½ hours on wing.

Unfortunately for Ron in 1987 builders erected several new bungalows on the site just to the rear of his loft, which did not help the birds in any way, so he decided to erect a new loft at the rear of his garden. His North Road YBs were raced to this loft, then after OB racing ended he re-settled them to their new home, and everything has been fine since.

Amongst his 1988 wins was 7th Club Lerwick, in 1989 2nd Club Lerwick, and the 1990 season proved to be a very good one as many prizes were won. Third highest prizewinner in club, winning three 1sts,

Ron Laws on loft, holding his 1992 Lerwick winner 'Little Gem 2', with helper Charlie Coulson on right holding the second Lerwick 'Little Gem 3'. Only two birds home on day in club, over 535 miles.

'LITTLE GEM 4'
Three year old blue hen of Mr & Mrs M Bishop bloodlines. Third Chatteris DHS, 11th SE Sect Fed, 23rd Sect F NRCC Lerwick.

Bubwith, Thurso and Berwick YB. In the Scottish races he took 10th Perth, 5th Fraserburgh; 5th Thurso (1), 3rd, 5th & 9th Lerwick; also 1st & 4th Thurso (2), winning eight trophies, and in the SE Sect of Fed he took 9th Sect Lerwick and 8th Sect Thurso (2), also winning diplomas in the Cambs CC.

The 1991 season proved to be an even better one for him, as he was second highest prizewinner, also winning 20 diplomas, ten specials, 13 trophies and five replicas. He won four races, Northallerton (1) OB, New Pitsligo OB, Thurso OB, and Berwick YB, and in those races he took 1st & 9th New Pitsligo, 6th & 10th Lerwick, 1st, 4th & 5th Thurso, also 1st, 2nd, 3rd & 4th Berwick YB, a very hard race when only eight birds

were timed in club on day from 245 miles. The main trophies won were
Bubwith to Berwick OB Av, New Pitsligo, Thurso, all Scottish averages,
r/u YB Av, and Combined Av, indeed a fine achievement. In that final
YB race from Berwick he won 3rd, 8th & 11th in the SE Sect of the
Peterborough & Dist Fed, and £54. He also won cash and diplomas in
the renowned Cambs CC.

1992 was another successful season, winning 20 diplomas, four specials,
ten trophies and four replicas. His best efforts being 1st, 2nd, 3rd, 7th
& 9th Lerwick (having the only two birds home in club on day), 1st &
5th Perth (2), 2nd, 3rd, 6th & 8th Thurso, also 1st Berwick YB (for the
third successive year). His trophies include Lerwick (two), Perth (two),
OB Av, Scottish Av, Lerwick & Thurso Av, Fraserburgh-Thurso &
Lerwick Av, Berwick YB and Combined averages. He took 5th SE Sect,
19th Fed Wetherby, 8th SE Sect Fraserburgh, 1st, 7th & 11th SE Sect,
and 11th Fed Lerwick, with 1,259 birds, 5th SE Sect Morpeth, 4th SE
Sect Perth (2) and 12th SE Sect Thurso, 447 miles.

In the NRCC Lerwick Classic he won 1st, 12th & 23rd Sect F and
121st Open, winning £230 cash, and from NRCC Thurso he won 26th
& 31st Sect F and 236th & 248th Open with 4,600 birds, and £50. His
first arrival from the Shetlands over 535 miles was a four year old blue
hen, now named 'Little Gem 2' being of Delbar and S B Cooper
bloodlines, a daughter of 'Little Gem' and 'Charlie's Boy' and a winner
of 1st Salisbury on South. She had flown Lerwick last year, then was
given two races from Morpeth this year prior to going to the Shetlands
where she was sent sitting ten days, to win the red card, etc. His second
arrival was another blue hen bred in 1988, 'Little Gem 3', of Delbar and
S B Cooper lines, being a grand-daughter of 'Little Gem' and she was
6th Club Lerwick on day last year. A very consistent performer, she was
also given two races from Morpeth then sent sitting ten days for Lerwick.

His first next morning pigeon was a blue hen bred in 1989, 'Litle Gem
4' of Mr & Mrs M Bishop of March bloodlines, a winner of other awards.
Flown Thurso last year, then given two races from Morpeth this year,
and sent to Lerwick sitting ten days. The winner of 7th Club Lerwick
was a blue four year old cock, same way bred as the race winner 'Little
Gem 2', being 11th Club Lerwick last year, and then sent to Thurso taking
6th Club, yet another gallant pigeon. The winner of 9th Club Lerwick
was a three year old chequer cock, similarly bred to the others.

Ron's Perth (2) winner is a yearling blue cock of Louella Janssen lines,
and this same pigeon won the hard Berwick YB race last year. His Berwick
YB winner this year was a black Krauth cock of Louella-Pick lines which
has flown consistently for him. One of his best distance pigeons in recent
years was a black seven year old hen, a winner in 1990 of 1st Thurso,
and in 1991 a winner of 1st New Pitsligo, 367 miles, and 1st Thurso, 447
miles.

The OB loft measures some 30ft in length, and there is a stock loft
and aviary plus a YB loft, with approx 30 pairs being wintered. This

year 50 youngsters were reared. Being interested in the distance races, nothing gives him greater pleasure than to see a bird arrive in good time from the Shetlands. Although he has tried a few birds on the south route, plus a few on Widowhood, it's all North Road racing now, and on the Natural system, saying "that is the best way for long distance racing". Present strains are mainly Krauth, Warrington, Delbar, S B Cooper, Vanhee, Keeble, and Westcott! All good long distance bloodlines.

A good sound mixture is fed all year round, and when mating mid-February it's mainly a case of putting "best to best". He is full of praise for fellow club member Fred Joyce for letting him have some of his best bloodlines, when he quit the Fancy a number of years ago (he has since returned). He also thanks his good friend and near neighbour Charlie Coulson who helps out with the chores, and will time in also in his absence. Another good friend Dennis Perry who takes the trainers.

Ron and his brother David also have a pet shop, for foods etc, and when training commences, Ron provides a van for this, which is available to any of the other club members if they care to take advantage of this fine gesture, at reasonable costs. Earlier in the year I handled most of Ron's birds, and realised that this was a loft of strength, and would take a lot of beating, and as he is coming up to 64 years of age, it won't be too long now before retirement comes, so that he can look after his birds full-time, and I'm certain he is going to take a lot of beating.

FROM ORANGE BOX TO THE TOP

by B D Jones Bros & Son, Blaenavon
1st Welsh Grand National New Pitsligo

We would firstly like to thank the Editor of Squills for asking us to write this article with the novice in mind. You must accept that pigeons will be lost during racing, training, from hawks, wires, shooting and weather conditions.

When joining a local flying club attend all meetings, become involved with the administration of the club, get to know what is going on and don't be afraid to have a say in how the club is run as new ideas often improve the running of the club.

Take part in the work involved, the setting and checking of race clocks, ringing and basketing the birds on race night.

There are three golden rules for all novice and established fliers to follow:

1. Never handle your own birds after being race rung.
2. Never handle your own pool sheet after it has been race marked by the officials.
3. Never set or check your own clock.

Never do any of these three as this could possibly lead to disqualification. The breaking of any of the three rules above could ruin any possible appeal over a race result specially any outstanding result in a Fed or National race.

After attending all meetings accept the conclusions arrived at and don't fall into the trap of taking things said personally. Learn to set and check clocks, a club cannot have too many members involved on race days.

How would I start setting up a good loft of birds? The loft itself should be well ventilated and large enough to accommodate the number of birds you are going to keep. Personally I use a stock section of 8ft deep by 7ft wide to accommodate about 12 pairs.

Stocking the loft. There are a number of ways of obtaining stock birds at a reasonable price. Beginners don't have to pay large amounts of money in setting up. Charity sales are a very good starting place to purchase birds, because of the nature of the sale local and outside fliers will usually give of their best birds. These fliers would not be party to giving less than their best. Along similar lines are the club sales. Many of the large studs hold autumn sales and half-price sales clearing the lofts for the coming year. There are good and bad birds in all breeds. Only experience will help you obtain birds that suit your methods of flying.

The pigeons we ourselves buy in we never race. We buy for bloodlines. these birds are for breeding only. Racing these pigeons may result in loss and you are back to square one. Hold on to your stock birds because they are the future of any success coming your way. We also take young birds from the stock birds to continue the bloodlines plus retiring the

The "orange box" loft where it all began in 1948.

H Webley presenting Brian and David with their first Osman Memorial
Trophy 1985.

David Holding, the Welsh Grand National Trophy for Thurso. Brian holding the Osman Memorial Trophy won for the second time.

very best race birds to the stock loft for the same reason.

Breeding

Normally we pair our stock birds and yearlings on the nearest Saturday to February 14 providing the weather is reasonable. Before the actual pairing takes place many hours have gone into pairing on paper. We take into account bloodlines and success at previous pairings.

After pairing the birds are secured in the nest box until the second egg is laid, this guarantees the parentage of the youngsters. 90% of our young birds are bred from the stock birds; the rest are taken off the race birds. Only the best race birds are allowed to breed youngsters plus a couple from the yearlings to give us an idea what stock birds to pair together in the future with the bloodlines in mind.

We commence our training about three weeks prior to the season. We begin with 20- to 30-mile tosses with exercise in the morning and evening for an hour. We like all our birds to have around five hours on wing approx 250- to 300-mile race for the longer races. The birds selected for key races and this takes place in the winter months, are then treated as individuals. Depending on the previous race conditions, they may race through to the big race date or be stopped racing some time before the big race. We like our hens to be sent sitting on ten- to 14-day old eggs on the day of basketing and the cocks to be sitting on 16-day old eggs or to a youngster 12 to 14 days old for the long distance races.

This is how we like to send our birds to the race but it does not always

The National winner.

work to plan so be ready to compromise. Remember in distance races plans don't always come off, so don't blame other people (eg) convoyer etc. The first person to blame is one-self and ask: "What have I done wrong?" and remember the lesson for next year.

With young birds we allow them to fly and roam the valleys to give them good education prior to commencing the training tosses. The youngsters start at one mile progressing up to 40 miles before their first race. All our youngsters are flown to the perch and not allowed to pair up. Our young birds race from 60 miles up to 280 miles.

Feeding

All our stock birds are fed the same corn all year round. The old and young birds are fed the same mixture as the stock birds except during racing when the old bird corn is altered for the long distance races. The basis of mix is 50% beans, 20% peas, 20% maize and 10% wheat; the beans, peas and wheat are bought from the farm. The beans are kept for two to three years before being used but sometimes used straight away from the farm for the young feed with no problem. The peas and wheat are used the following season, but the maize is bought from the corn merchant. So the price of our corn per cwt is £10.50 compared to around £20 in the corn merchant.

This is one way of cutting the cost down for the novice by buying from the farm but watch the beans are dried and watch there are no cracks in the beans. If there are then leave them there as they have been dried too quickly. The only thing we use in the water is multivitamins, once a week for the old birds and twice a week for the young birds. We believe simple methods are the best for long distance races and also quietness to keep the birds happy in the loft.

In 1992 we won our second National, this time from new Pitsligo. This was won by a cock 'Church Top Boy'. He won 1st Club, 1st Open New North Road Fed, 1st Fed NNRF, 1st Welsh Championship Club, 1st East Sect Welsh Grand National, 1st Open Welsh Grand National, 1st East Sect Welsh Combine and 1st Open Combine. He has also gained 5th Shrewsbury (2) YB, 1st Club, 16th Fed NNRF Northallerton; 4th Club Pontefract; 6th Club Thurso, two weeks after winning the National.

In conclusion

We would like to thank the members of Blaenavon Pigeon Club for their sportsmanship in racing in the club through the years. Also P Ingles for the sire of our National winner. The dam was bred by Louella Pigeon World and so was the grandsire. Our thanks to the Massarella family for the gift birds over the years and to date a mate to go with the National winner.

We hope this will give the novice the hope that they can beat the top fliers in their area without paying large amounts of money for pigeons, because this is the only sport where the schoolboy or girl can compete with the millionaire on equal terms. Good luck for the future.

DICK'S DELIGHT

by Dick Rumph & Son
1st Welsh Grand National Flying Club Thurso
1st Welsh Combine Thurso, 480 miles

It was indeed a great honour and privilege to receive an invitation from the Editor of Squills to contribute an article for the 1993 Squills Year Book. We hope the following few lines will be of some help to aspiring fanciers everywhere.

Let me start by first thanking all those kind people who offered me congratulations, either by letter or phone calls, when they heard of our success in the 1992 Thurso National race. Their kind thoughts were greatly appreciated.

I first entered the sport in 1969 and my initial contact was made due to regular visits to the loft of my late brother Ronnie, who kindly started me off on the road to success by gifting to me a few birds. One of these was a stock cock which Ron had obtained from the late Tom Pike of Maerdy in the Rhondda. This cock was to prove to be the foundation of our loft as he was responsible for birds to win from the first race Craven Arms, 60 miles, through to the longest race Lerwick, 596 miles. Over

Dickie Rumph & Son, 1st WGNFC Thurso 1992, pictured outside loft with the National winner and his sire. Photo by John Dixon.

the years these birds have enabled us to top the West Section on three separate occasions plus many more positions besides.

Our National winner 'Dick's Delight' was bred from a mealy cock which himself has been an exceptional racer for us, winning seven 1sts including 1st Thurso, 480 miles. The dam of 'Dick's Delight' was presented to us by my nephew Allan Rumph of Caerau. The local federation in which we fly our birds weekly is the South West Glamorgan Federation which we have topped on many occasions, also winning many Open races.

Our feeding method is quite simple as we feed Willsbridge Number 1 Racing Mixture which consists of mature beans, peas and tares and the birds are allowed to eat their fill as we believe in feeding the birds well and working them hard. We also feed Red Band conditioning seed as a titbit. The birds are trained very hard and must earn their perch.

In closing our advice to novices in the sport is to have plenty of patience and success will surely follow as it has for us.

JANUARY

1 FRI	NEW YEAR HOLIDAY
2 SAT	
3 SUN	
4 MON	BANK HOLIDAY (SCOTLAND)
5 TUES	
6 WED	
7 THURS	
8 FRI	
9 SAT	
10 SUN	
11 MON	
12 TUES	
13 WED	
14 THURS	
15 FRI	

NOTES

16 SAT		24 SUN	
17 SUN		25 MON	
18 MON		26 TUES	
19 TUES		27 WED	
20 WED		28 THURS	
21 THURS		29 FRI	
22 FRI		30 SAT	
23 SAT		31 SUN	

NOTES

FEBRUARY

1 MON	
2 TUES	
3 WED	
4 THURS	
5 FRI	
6 SAT	
7 SUN	

8 MON	
9 TUES	
10 WED	
11 THURS	
12 FRI	
13 SAT	
14 SUN	ST VALENTINE'S DAY
15 MON	

NOTES

16 TUES	
17 WED	
18 THURS	
19 FRI	
20 SAT	
21 SUN	
22 MON	
23 TUES	

24 WED	
25 THURS	
26 FRI	
27 SAT	
28 SUN	

NOTES

MARCH

1 MON	ST DAVID'S DAY (WALES)
2 TUES	
3 WED	
4 THURS	
5 FRI	
6 SAT	
7 SUN	

8 MON	
9 TUES	
10 WED	
11 THURS	
12 FRI	
13 SAT	
14 SUN	
15 MON	

NOTES

16 TUES	
17 WED	ST PATRICK'S DAY (IRELAND)
18 THURS	
19 FRI	
20 SAT	
21 SUN	
22 MON	
23 TUES	

24 WED	
25 THURS	
26 FRI	
27 SAT	
28 SUN	
29 MON	
30 TUES	
31 WED	

NOTES

APRIL

1 THURS	
2 FRI	
3 SAT	
4 SUN	
5 MON	
6 TUES	PASSOVER
7 WED	
8 THURS	
9 FRI	GOOD FRIDAY
10 SAT	
11 SUN	EASTER SUNDAY
12 MON	EASTER MONDAY
13 TUES	
14 WED	
15 THURS	

NOTES

16 FRI	
17 SAT	
18 SUN	
19 MON	
20 TUES	
21 WED	
22 THURS	
23 FRI	ST GEORGE'S DAY (ENGLAND)

24 SAT	
25 SUN	
26 MON	
27 TUES	
28 WED	
29 THURS	
30 FRI	

NOTES

MAY

1 SAT	
2 SUN	
3 MON	MAY DAY HOLIDAY
4 TUES	
5 WED	
6 THURS	
7 FRI	

8 SAT	
9 SUN	
10 MON	
11 TUES	
12 WED	
13 THURS	
14 FRI	
15 SAT	

NOTES

16 SUN		**24** MON	
17 MON		**25** TUES	
18 TUES		**26** WED	
19 WED		**27** THURS	
20 THURS		**28** FRI	
21 FRI		**29** SAT	
22 SAT		**30** SUN	
23 SUN		**31** MON	SPRING BANK HOLIDAY

NOTES

JUNE

1 TUES	
2 WED	
3 THURS	
4 FRI	
5 SAT	
6 SUN	
7 MON	HOLIDAY (EIRE)

8 TUES	
9 WED	
10 THURS	
11 FRI	
12 SAT	
13 SUN	
14 MON	
15 TUES	

NOTES

16 WED		24 THURS	
17 THURS		25 FRI	
18 FRI		26 SAT	
19 SAT		27 SUN	
20 SUN		28 MON	
21 MON	LONGEST DAY	29 TUES	
22 TUES		30 WED	
23 WED			

NOTES

JULY

1 THURS	
2 FRI	
3 SAT	
4 SUN	
5 MON	
6 TUES	
7 WED	

8 THURS	
9 FRI	
10 SAT	
11 SUN	
12 MON	HOLIDAY (N. IRELAND)
13 TUES	
14 WED	
15 THURS	

NOTES

16 FRI		24 SAT	
17 SAT		25 SUN	
18 SUN		26 MON	
19 MON		27 TUES	
20 TUES		28 WED	
21 WED		29 THURS	
22 THURS		30 FRI	
23 FRI		31 SAT	

NOTES

AUGUST

1 SUN	
2 MON	BANK HOLIDAY (Scotland/Eire)
3 TUES	
4 WED	
5 THURS	
6 FRI	
7 SAT	
8 SUN	
9 MON	
10 TUES	
11 WED	
12 THURS	
13 FRI	
14 SAT	
15 SUN	

NOTES

16 MON		24 TUES	
17 TUES		25 WED	
18 WED		26 THURS	
19 THURS		27 FRI	
20 FRI		28 SAT	
21 SAT		29 SUN	
22 SUN		30 MON	SUMMER BANK HOLIDAY
23 MON		31 TUES	

NOTES

SEPTEMBER

1 WED	
2 THURS	
3 FRI	CLOSING DATE REDUCED RATE ADVERTS SQUILLS 1994
4 SAT	
5 SUN	
6 MON	
7 TUES	
8 WED	
9 THURS	
10 FRI	
11 SAT	
12 SUN	
13 MON	
14 TUES	
15 WED	

NOTES

16 THURS		24 FRI	
17 FRI		25 SAT	
18 SAT		26 SUN	
19 SUN		27 MON	
20 MON		28 TUES	
21 TUES		29 WED	
22 WED		30 THURS	
23 THURS			

NOTES

OCTOBER

1 FRI	
2 SAT	
3 SUN	
4 MON	
5 TUES	CLOSING DATE SURCHARGED ADVERTS SQUILLS 1994
6 WED	
7 THURS	

8 FRI	
9 SAT	
10 SUN	
11 MON	
12 TUES	
13 WED	
14 THURS	
15 FRI	

NOTES

16 SAT		24 SUN	
17 SUN		25 MON	
18 MON		26 TUES	
19 TUES		27 WED	
20 WED		28 THURS	
21 THURS		29 FRI	HOLIDAY (EIRE)
22 FRI		30 SAT	
23 SAT		31 SUN	

NOTES

NOVEMBER

1 MON	
2 TUES	
3 WED	
4 THURS	
5 FRI	
6 SAT	
7 SUN	
8 MON	
9 TUES	
10 WED	
11 THURS	REMEMBERANCE DAY
12 FRI	
13 SAT	
14 SUN	
15 MON	

NOTES

16 TUES		24 WED	
17 WED		25 THURS	
18 THURS		26 FRI	
19 FRI		27 SAT	
20 SAT		28 SUN	
21 SUN		29 MON	
22 MON		30 TUES	ST ANDREWS DAY (SCOTLAND)
23 TUES			

NOTES

DECEMBER

Date	
1 WED	
2 THURS	
3 FRI	
4 SAT	
5 SUN	
6 MON	
7 TUES	
8 WED	
9 THURS	
10 FRI	
11 SAT	BRITISH NATIONAL (OLD COMRADES) SHOW
12 SUN	BRITISH NATIONAL (OLD COMRADES) SHOW
13 MON	
14 TUES	
15 WED	

NOTES

16 THURS		24 FRI	
17 FRI		25 SAT	CHRISTMAS DAY
18 SAT		26 SUN	BOXING DAY
19 SUN		27 MON	BANK HOLIDAY MONDAY
20 MON		28 TUES	BANK HOLIDAY TUESDAY
21 TUES		29 WED	
22 WED		30 THURS	
23 THURS		31 FRI	

NOTES

JANUARY 1994

Date	
1 SAT	
2 SUN	
3 MON	NEW YEAR BANK HOLIDAY
4 TUES	BANK HOLIDAY (SCOTLAND)
5 WED	
6 THURS	
7 FRI	
8 SAT	
9 SUN	
10 MON	
11 TUES	
12 WED	
13 THURS	
14 FRI	
15 SAT	

NOTES

16 SUN		24 MON	
17 MON		25 TUES	
18 TUES		26 WED	
19 WED		27 THURS	
20 THURS		28 FRI	
21 FRI		29 SAT	
22 SAT		30 SUN	
23 SUN		31 MON	

NOTES

1994 Year Planner

Wk.		Mon	Tues	Wed	Thurs	Fri	Sat	Sun
1		27	28	29	30	31	**1**	2
2	JANUARY	3	4	5	6	7	8	9
3		10	11	12	13	14	15	16
4		17	18	19	20	21	22	23
5		24	25	26	27	28	29	30
6		31	**1**	2	3	4	5	6
7	FEBRUARY	7	8	9	10	11	12	13
8		14	15	16	17	18	19	20
9		21	22	23	24	25	26	27
10		28	**1**	2	3	4	5	6
11	MARCH	7	8	9	10	11	12	13
12		14	15	16	17	18	19	20
13		21	22	23	24	25	26	27
14		28	29	30	31	**1**	2	3
15	APRIL	4	5	6	7	8	9	10
16		11	12	13	14	15	16	17
17		18	19	20	21	22	23	24
18		25	26	27	28	29	30	**1**
19		2	3	4	5	6	7	8
20	MAY	9	10	11	12	13	14	15
21		16	17	18	19	20	21	22
22		23	24	25	26	27	28	29
23		30	31	**1**	2	3	4	5
24	JUNE	6	7	8	9	10	11	12
25		13	14	15	16	17	18	19
26		20	21	22	23	24	25	26
27	JULY	27	28	29	30	**1**	2	3
28		4	5	6	7	8	9	10

Mon	Tues	Wed	Thurs	Fri	Sat	Sun		Wk.
11	12	13	14	15	16	17	JULY	29
18	19	20	21	22	23	24		30
25	26	27	28	29	30	31		31
1	2	3	4	5	6	7	AUGUST	32
8	9	10	11	12	13	14		33
15	16	17	18	19	20	21		34
22	23	24	25	26	27	28		35
28	30	31	**1**	2	3	4	SEPTEMBER	36
5	6	7	8	9	10	11		37
12	13	14	15	16	17	18		38
19	20	21	22	23	24	25		39
26	27	28	29	30	**1**	2	OCTOBER	40
3	4	5	6	7	8	9		41
10	11	12	13	14	15	16		42
17	18	19	20	21	22	23		43
24	25	26	27	28	29	30		44
31	**1**	2	3	4	5	6	NOVEMBER	45
7	8	9	10	11	12	13		46
14	15	16	17	18	19	20		47
21	22	23	24	25	26	27		48
28	29	30	**1**	2	3	4	DECEMBER	49
5	6	7	8	9	10	11		50
12	13	14	15	16	17	18		51
19	20	21	22	23	24	25		52
26	27	28	29	30	31	**1**	JANUARY 1995	1
2	3	4	5	6	7	8		2
9	10	11	12	13	14	15		3
16	17	18	19	20	21	22		4

PERSONAL & LOFT INFORMATION

Name & Address...

....

....

Union or Association Membership Code No ....

National or Specialist Club Code No ....

CLOCK NUMBERS

Make...............................Serial No................................

Make...............................Serial No................................

Make...............................Serial No................................

CLUB SECRETARIES

1. ....

....................................Phone......................................

2. ....

....................................Phone......................................

3. ....

....................................Phone......................................

FEDERATION OR COMBINE SECRETARIES

1. ....

....................................Phone......................................

2. ....

....................................Phone......................................

Loft Number ....

Latitude...North

Longitude..East or West

SQUILLS REFERENCE SECTION

RECORD PERFORMANCES
CLASSIC WINNERS
    NATIONAL FC
    NORTH ROAD CC
    SCOTTISH NATIONAL FC
    BRITISH INTERNATIONAL CC
    WELSH GRAND NATIONAL FC
    BRITISH BARCELONA CLUB
    IRISH NATIONAL FC
    UP NORTH COMBINE
FEDERATIONS AFFILIATED TO FCI
OLYMPIADE WINNERS
INTERNATIONAL JUDGING STANDARD
COLOUR ABBREVIATIONS
PAIRS FOR COLOUR BREEDING
NATURAL RACING TABLE
TIME OF FLIGHT APPROXIMATOR (YPM)
METRIC CONVERSION TABLES
METRIC TIME OF FLIGHT APPROXIMATOR
COMPARATIVE VALUES OF FOODS
OSMAN MEMORIAL CUP WINNERS
WHERE TO REPORT FOUND BIRDS

RECORD PERFORMANCES

All Velocities in Yards per Minute

UP TO 250 MILES

40 Miles (approx)—E Croydon Race, E Anglia Fed, 1965. A Vidgeon & Son, Wickford, vel 3229 (110 mph)

80 Miles—Malahide Race, Ulster Fed, 1914. H Mussen, vel 2744

88 Miles—Ed Jalowiec, Lakewood, 1983, vel 2892

100 Miles—Guernsey Race, Dorchester Club, 1975. L Holbrook, Newport (IoW), vel 2385

100 Miles—Waiuoru Race, Masterton Club, New Zealand, 1962. L Groube, vel 2578

139 Miles—Isohara Race, Kanagawa Fed (Japan), 1960. S Ichifujiki, vel 2618

150 Miles—Christchurch Race, 1962. Dyke & Hayward, Lowestoft, vel 2362

162 Miles—Wexford Race, Glengormley HPS, 1964. T Farlowe, vel 2364

186 Miles—Dungarvan Race, Alexandra HPS, Ulster Fed, 1961. Beattie Bros, vel 2857

200 Miles—Maple City PC (USA), 1962. B C Mahrle, vel 2088 (2 birds)

203 Miles—Stuart Child, Oklahoma YB, 1979, vel 2294

210 Miles—Glen Race, Southern Suburbs HS, Transvaal Fed, 1945. T Wood, vel 2256

215 Miles—Pete van de Merwe, Utah, 1971, vel 2435

250 TO 500 MILES

268 Miles—Nou Race, Akita Fed (Japan), 1966. A Sawaguchi, vel 2078

274 Miles—J & P Jones, Texas, 1982, vel 2183

288 Miles—Kohma Race, Saitama Fed (Japan), 1969. S Komiya, vel 1827

289 Miles—Stuart Child, Oklahoma YB, 1974, vel 2256

304 Miles—W Kaczmarek, Illinois YB, 1981, vel 2245

305 Miles—Melvin Walther, Texas, 1982, vel 2243

346 Miles—Fraserburgh Race, Coalville Central HS, 1968. B Locke, vel 2032

370 Miles—Marree Race, S Australia HPA, 1947. C Copelands, vel 2825

390 Miles—Horrell Simpson, Texas, 1984, vel 2340

401 Miles—P Bingham, Texas, 1984, vel 2310

403 Miles—Maquassi Race, Beaufort West HS (South Africa), 1954. C Nolte, vel 1875

436 Miles—Nantes, Warrington 2-B Club, 1952. H Whittle, vel 1764

441 Miles—Wooramel Race, Canning Club (W Australia), 1974. T Humphreys, vel 2375

455 Miles—Al Kryl, Lawrenceville YB, 1965, vel 2065

465 Miles—Al Kryl, Lawrenceville YB, 1965, vel 2065

474 Miles—Hillsport Race, Far North RPC, Canada, 1963. H Meuhling, vel 1583

478 Miles—SNFC Rennes Race, 1967. R Hope, vel 1645

500 TO 750 MILES

500 Miles (2 birds)—Knightstown Race, Susquehanna Centre (USA), 1957. Rabzak Bros, Reading, vels 1993 and 1988

500 Miles—Thurso, London NR Combine, 1948. W Reed, vel 2095

501 Miles—R V Smith, South Africa, 1979, vel 2583 (88 mph)

534 Miles—Rennes, Scottish National, 1972. R & H Kennedy, Irvine, vel 1587

535 Miles—Frank Vrzak, Nebraska, 1968, vel 2159

539 Miles—Rennes, Scottish National, 1967. J Law, Easthouses, vel 1503

545 Miles—Oodnadatta Race, PRPC (S Australia), 1973. M Lever, vel 1791.60

547 Miles—Rennes, 1st Scottish National FC, 1923. White & Dickson, Paisley, vel 1488

562 Miles—Hanover, British Berlin FC, 1952.

Hollingsworth & Edwards, vel 1658

586 Miles—Aycock Family, Arizona, 1984, vel 1809

600 Miles—Sandusky Race, Greater Boston Concourse, 1957. Morris Gordon, vel 2000

601 Miles—Lerwick, 2nd prize, North Road CC, 1931. Dr Tresidder, London, vel 1684

621 Miles—Joe Robinson, Texas, 1952, vel 2070

636 Miles—San Sebastian, Harborough and District HS, 1907. J Wones, Wombourne. (Liberated 11.55am, July 22; timed in 4.54pm, July 23)

638 Miles—Bombay, India Race to Madras. Mr Baldrey's vel 750

645 Miles (on day)—Up North Combine, Angouleme, 1958. J Latimer, Whitburn, vel 1192

653 Miles—Wakkanai Grand National, Takaoka Fed (Japan), 1975. R Ibayashi, vel 1537

654 Miles (on day)—West of England Lerwick Club, Lerwick Race 1978. C Foster, Dorchester, vel 1603

660 Miles—Matjesfontein Race, Pretoria RPC, Transvaal Fed, 1951. J J Bouwer, vel 1534

663 Miles (on day)—Scottish National, Rennes, 1967. P McDonald, Fraserburgh, vel 1178

676 Miles—Stuart Child, Oklahoma, 1973, vel 1881

680 Miles—Mr & Mrs Nash, Southampton, London Columbarian Barcelona 1963, vel 680

686 Miles—Hanover, British Berlin FC, 1952. A R Hill, vel 1309

692 Miles—Nevers, 1st Scottish Central Combine, 1929. R Stoddart, Carron, vel 959

696 Miles—Faroe Isles Race, 1929. G F Baker, Peterborough, vel 1262

706 Miles—J B Selby, Oklahoma, 1973, vel 1860

710 Miles (on day)—Hicks Bay Race, Dunedin Club, New Zealand. J Still & C Watson

710 Miles—Up North Combine, Le Puy, 1934. R Jones, Shotton Colliery, vel 1006

723 Miles (on day)—Nantes, Scottish National FC, 1977. R Whyte, Fraserburgh, vel 1594

724 Miles—Nantes. E Cardno, Fraserburgh, vel 1648

727 Miles—Mirande National, 1909. Dr Barker's 'True Blue', vel 839

737 Miles—Wakkanai Grand National, Fukui Fed (Japan), 1975. S Sumino, vel 1433.72

740 Miles (on day)—Bassanno, Alberta Race, N Winnipeg FC, Canada, 1945. J Rowski, 1348; Wilkes, 1345.9; Burgess, 1345.6

750 Miles—Belville, Cape Town Race, Rand Central HS, Transvaal Fed, 1951. Whelehan, vel 1261.1

OVER 750 MILES

751 Miles (on the day)—Lulea Race, 1951. Liberated 1am, flying time 15 hours 52 minutes to Ystrad, Sweden. Edfelt & Wahlstrom, vel 1400 ypm (approx)

756 Miles—Pau NFC, 1976. M Purdom, Carlisle, vel 1095

758 Miles—Pau Race, National FC, 1950. T Dixon, Askam, vel 1044

762 Miles—Faroe Isles Race, NRCC, 1929. Fuller-Isaacson, London, vel 1029

764 Miles—San Sebastian Race, NFC, 1934. J T Clark, Windermere, vel 909

788 Miles—Marseilles Open Race, 1927. F E Brown, Bridlington, vel 939

789 Miles—Eddontenajon Race, Columbia RPC (Canada) 1973. George Shanks, vel 1115.9 and 1006.3

791 Miles—Eddontenajon Race, Columbia RPC (Canada) 1973. George Staudinger, vel 1113 and 1047.2 (see also 789.7 miles)

794 Miles—E R Shockey, Texas, 1979, vel 1603

803 Miles—San Antonio Club, Texas, USA, 1941. E S Petersen, vel 1525

829 Miles 1400 yards—Mariental Race, July 1969. A C Duff, vel 1430

830 Miles—Salisbury Race (South Africa), 1949. Scotney & Son, vel 1112

834 Miles—Daniels Bros, Grappenhall, British Barcelona Club, Palamos, 1968, vel 856

854 Miles—Bordeaux to Aberdeen. G W Henson in the 1902 National

861 Miles—Palamos BBC, 1971. J & J G Paley, Silsden, vel 788

864 Miles—Palamos BBC, 1991. R G Wales, Malton, N Yorks, vel 965

870 Miles—San Sebastian Race, NFC, 1934. A Millar, Wishaw. Liberated 13 July; homed 21 July

870 Miles—Fukue Grand National, Aomori Fed (Japan), 1972. Ho Ogasawara, vel 1716.55

873 Miles—Barcelona International Race, 1961. J C Tattersall, Bradford, vel 270

888 Miles—San Sebastian Race. Swain Bros, Markinch. Liberated 13 July; homed 1 August

911 Miles—All Japan Champion, Kure Fed, 1974. H Oda, vel 976

919 Miles—Barcelona Race, 1931. T Dixon, Askam, 7½ days

938 Miles—Barcelona International, 1967. R P Eggleston, Long Marton, Westmorland, 3 days 2 hours 50 minutes, vel 511

953 Miles—Barton Race (South Australia), 1933. Liberated 18 October; arrived 30 October. T Brown, Dumbley. Pigeon was nine months olds

976 Miles—Barcelona BBC, Mr & Mrs Graham, Ecclefechan

1000 Miles (approx)—Salisbury Race, Queenstown Club, S Africa, 1950. Casey Bros, vel 1400

1001 Miles—Rome Race, 1913. C H Hudson, Derby. Liberated 4.30am 29 June; homed 1.15pm 29 July

1002 Miles—Fort Wayne, Indiana, USA, Club, 1927. C W Oetting, 1 day 10 hours 22 minutes

1004 Miles—Barton Race (S Australia), 1934. L Williams 26 hours 51 minutes 40 seconds flying time. 2nd E Hunt beaten by 2 seconds

1010 Miles—Palamos BBC. Alan Raeside, Irvine, 43 hours 56 minutes

1013 Miles—British Barcelona Club, Palamos Race, 1975. A Raeside Irvine, vel 675

1030 Miles—Hays, Kansas Race, 1934. H Elston, Glenfield Even, USA, 31 hours racing time

1033 Miles—Barcelona Race, 1960. W G Davidson, Stevenston, Ayrshire. 23 days 23 hours 50 minutes

1093 Miles—Rome to Spennymoor. Vester & Scurr. Liberated 4.30am 29 June, 1913; homed 8.44am 17 August

1094 Miles—H A Burdwood, Northeast, 1968, vel 1220

1107 Miles to Edmonton, Canada. E Thompson. Liberated 6am 14 July; timed 6pm 18 July 1939

1110 Miles—San Antonio Club, Texas, USA, 1948. E S Petersen, vel 445

1141 Miles—Barcelona Race, 1960. A Bruce, Fraserburgh. Liberated 9 July; homed 5 August

1173 Miles—Palamos, Spain to Wick, Scotland, 1976. Rosie & Bruce

1351 Miles—San Antonio Club, Texas, USA, 1946. E S Petersen, vel 372

1427 Miles—Great Falls Race, San Antonio Club, Texas, USA, 1953. E R Shockey, vel 1125

2039 Miles—Churchill, Manitoba, Canada, Race, San Antonio Club, Texas, USA, 1937. R W Taubert, 43 days 10 hours

5400 Miles—Ichabo Isle, W Africa to Nine Elms, London. (Dropped dead one mile from loft) 55 days, liberated 1 June 1845. Owned by Duke of Wellington. 5,400 miles is approximate airline distance, but as bird could not fly across the Sahara Desert the probable distance is about 7,000 miles

CLASSIC WINNERS

NATIONAL FLYING CLUB

1898 (Bordeaux)—J Ward, Oldham
1899 (Lerwick)—HRH the Prince of Wales, Sandringham
      (Bordeaux)—J W Toft, Liverpool
1900 (Bordeaux)—S H Shinner, Bromsgrove
1901 (Bordeaux)—Orchardson, Liverpool
1902 (Bordeaux)—J L Baker, Sedgley
1903 (La Roche)—Elsden, Brighton
1904 (Marennes)—Cannon, Pontypridd
1905 (Marennes)—Matten, Epsom
1906 (Marennes)—Stephens, Plymouth
1907 (Marennes)—Jenkins, Worcester
1908 (Mirande)—J Wones, Wombourne
1909 (Mirande)—W Saunders, Cradley
1910 (Mirande)—Higham Bros, Bradenstoke
1911 (Bordeaux)—Mitchell Bros, Dudley
1912 (Dax)—G W Scadden, Portsmouth
      (Bordeaux)—P Brough, Congleton
1913 (Bordeaux)—E H Grellett, Hitchin
1914 (Marennes)—Deakin & Redfern, Sheffield
      (Pons)—J Wones, Wombourne
      (Bordeaux)—H T Stratton, Bournemouth
1915 to 1919—No Races
1920 (Bordeaux)—S Theelan, Worth
1921 (San Sebastian)—W H Peters, Plumstead
1922 (San Sebastian)—J W Logan, East Langton
1923 (San Sebastian)—E A Turner, South Norwood
1924 (San Sebastian)—F W Marriott, Birmingham
1925 (San Sebastian)—F W Marriott, Birmingham
1926 (San Sebastian)—M Edmunds, Llantwit Fardre
1927 (San Sebastian)—W Ashman, Ogmore Vale
1928 (San Sebastian)—R Wright, Burscough
1929 (San Sebastian)—W B Reeve, Wickham Market
1930 (San Sebastian)—H F Hoole, Sutton
1931 (San Sebastian)—W R James, Wolverhampton
1932 (San Sebastian)—R M Antingham, Oxford
1933 (San Sebastian)—Elliott Bros, Brixton
1934 (San Sebastian)—V Robinson, Southampton
1935 (San Sebastian)—Sir William Jury, Reading
1936 (San Sebastian)—R Tustian, Great Tew
1937 (Mirande)—W H Hardcastle, Birmingham
1938 (Mirande)—F Marks, East Grinstead
1939 (Mirande)—G Weller, Ewhurst
1940 (Penzance)—F W Marriott, Birmingham
1941 (Penzance)—Lambert & Son, Mountain Ash
1942 (Penzance)—Parry & Corfield, Poole
1943 (Penzance)—J H Long, Birmingham
1944 (Penzance)—R F Boyen, Ilfracombe
1945 (Penzance)—Sharrock Bros, Haskayne
1946 (Bordeaux)—L Gilbert, Forest Hill
1947 (Bordeaux)—G E Jarvis, Eastry
1948 (Guernsey)—Fearn & Son, Wirksworth
1949 (Luxembourg)—F H Jarvis, Harpenden
1950 (Pau)—Moore & Wootton, Kidsgrove
1951 (Pau)—F Godfrey, Worksop
1952 (Pau)—G King, Hastings
1953 (San Sebastian)—Barker Bros, Hucknall
1954 (Pau)—Ranaboldo Bros, W Molesey
1955 (Pau)—V Robinson, Southampton
1956 (Pau)—J W Langstone, Worcester
1957 (Pau)—Skone & Everall, Shrewsbury
1958 (Bordeaux)—W Middle, Weston-super-Mare
1959 (Pau)—H J King, Camberley
1960 (Pau)—F E Naylor, Thelwall
1961 (Pau)—C Chalk, Bitterne, vel 940 ypm
      (Nantes)—P Bridgewater, Godalming, vel 1275 ypm
1962 (Nantes)—F Wensley, Bromsgrove, vel 1181 ypm
      (Pau)—R Mitcheison & Son, Winchester (1079)
1963 (Nantes)—R H Christopher, Fontwell Magna
      (Pau)—T Lancashire, Aylesham
1964 (Nantes)—L Lewis, Farnborough
1965 (Pau)—J O Warren & Son, Banks, Southport
1965 (Nantes)—R Peters, Frindsbury, Kent, vel 1319
      (Pau)—S J Banfield, Bournemouth, vel 1449.
1966 (Nantes)—L C Shannon & Son, Lands End, vel

1727
(Pau)—Willcox Bros, Clutton, Bristol, vel 937
1967 (Nantes)—W L Gale, Weymouth, vel 1095
(Pau)—J McLaren, Portsmouth, vel 922
1968 (Nantes)—K St Panzer, Eastleigh, vel 1289
(Pau)—J Adams, Redditch, vel 859
1969 (Nantes)—A H Bennett, Church Stretton, vel 1392
(Pau)—K Morey, E Cowes, IoW, vel 1015
1970 (Nantes)—C E Pollard & Son, Camborne, vel 1189
(Pau)—L Davenport, Ropley, vel 1224 ypm
1971 (Nantes)—Stevenson, Horndean, vel 1292 ypm
(Pau)—A T Jarvis & Son, Bridgwater, vel 865
1972 (Nantes)—M S J Lilliott, Ashford, vel 1312
(Pau)—L Adams, Bournemouth, vel 1016
1973 (Nantes)—R Higgs, Frimley Green, vel 1180
(Pau)—G Wood, Ventnor, vel 1151
1974 (Nantes)—Mr & Mrs O'Brien, Eastbourne, vel 1214
(Pau)—A Mark, Croydon, vel 941
1975 (Nantes)—G Burgess, Alderley Edge, vel 1401
(Pau)—L Davenport, Alresford, vel 1246
1976 (Nantes)—A T Jarvis & Son, Bridgwater, vel 1314
(Pau)—A Payne, Keynsham, vel 1288
1977 (Nantes)—P Thomas, Penzance, vel 1016
(Pau)—T Totterdale, Nailsworth, vel 819
1978 (Nantes)—B Corbin, Southampton, vel 1090
(Pau)—R Oldfield, Clanfield, vel 1120
1979 (Nantes)—K J Thorpe, Colchester, vel 1456
(Pau)—Fear Bros, Clandown, vel 933
1980 (Nantes)—D Harrison, St Helens, vel 1506
(Pau)—J Jackson, Worthing, vel 1059
1981 (Nantes)—B Thorpe & Son, Pinxton, vel 1409
(Pau)—Mr & Mrs Wasey, New Addington, vel 1263

1982 (Nantes)—B Wigginton, Netley Abbey, vel 1497
(Pau)—A H Bennett, Church Stretton, vel 972
1983 (Sennen Cove)—P Bartholomew, Brighton, vel 1865
1984 (Nantes)—Mr & Mrs A Bedford, Bristol, vel 1323
(Pau)—F Bloor, Charlton Mackrell, vel 1135
1985 (Nantes)—G W Kirkland, Telford, vel 1294
(Pau)—Mr & Mrs G Stovin, Didcot, vel 1306
1986 (Nantes)—A Mabin, Torquay, vel 1216
(Pau)—D & D Woodhouse Bros, Lake, IoW, vel 971
1987 (Nantes)—M King, Charlton Marshall, vel 1435
(Pau)—Miles & Dix, Peasedown, vel 1107
1988 (Nantes)—M King, Charlton Marshall, vel 1058
(Pau)—G Burgess, Wraysbury, vel 1244
1989 (Nantes)—E W Stoodley, Poole, vel 1666
(Pau)—D Delea, Rainham, vel 1046
1990 (Nantes)—P Hollard, Dorset, vel 1150.7
(Pau)—J McGee, W Sussex, vel 1200.3
1991 (Nantes)—M Spencer, Bacup, vel 1351
(Pau)—J Newall & son, Bedfont, vel 1230
1992 (Nantes)—L Nichols, Bedminster, vel 1424
(Pau)—P Kendal, Wantage, vel 930
(Saintes)—A Winstanley, Congleton, vel 1359

NORTH ROAD CHAMPIONSHIP CLUB

1901 (Lerwick)—J T Hincks, Leicester
1902 (Lerwick)—G Pulley, London
1903 (Lerwick)—H Pickering, Newhall
1904 (Lerwick)—P Clutterbuck, Sarratt
1905 (Thurso)—C Clark, Rushden
1906 (Thurso)—T Gill, Summers Bridge
1907 (Lerwick)—R Mattock, Southgate
1908 (Lerwick)—A Braithwaite, Leeds
1909 (Lerwick)—A Crawford, Pickering
1910 (Lerwick)—Hubbard & Warren, Nottingham
1911 (Lerwick)—D Fitzjohn, Peterborough
1912 (Lerwick)—Hustwaite Bros, Wellingborough
(Faroe)—T Long, Harrow
1913 (Lerwick)—W L Thackray, Old Malton
1914 (Lerwick)—P Clutterback, Sarratt
1915 to 1918—No Races
1919 (Thurso)—HM King George V, Sandringham
1920 (Lerwick)—F W Marriott, Saltley
1921 (Lerwick)—F W Marriott, Saltley
1922 (Lerwick)—C Clark, Rushden
1923 (Lerwick)—G W Gammons, Huntingdon

1924 (Lerwick)—J S Hartridge, Leicester
1925 (Lerwick)—Brown & Harton, Oakham
1926 (Lerwick)—W Westcott, Ipswich
1927 (Lerwick)—B J Westcott, Ipswich
1928 (Lerwick)—Captain Quibell, Newark-on-Trent
1929 (Lerwick)—Brown & Harton, Oakham
(Faroe)—G F Baker, Peterborough
1930 (Lerwick)—J S Hartridge, Leicester
(Faroe)—J S Hartridge, Leicester
1931 (Lerwick)—G W Graves, Newark-on-Trent
(Faroe)—Captain Quibell, Newark-on-Trent
1932 (Lerwick)—F Stevenson, Skegness
1933 (Lerwick)—Branstone, Son & Smith, Mansfield
1934 (Lerwick)—S R Atkins, Ipswich
1935 (Lerwick)—J R Marriott, Derby
1936 (Lerwick)—D Robotham & Sons, Leicester
1937 (Lerwick)—R F Towle, Totton
(Arendal)—J Banks, Folkestone
1938 (Lerwick)—Major G S Rogers, Sheringham
1939 (Lerwick)—Jepson Bros & Curtis, Peterborough
1940 (Fraserburgh)—North & Son, Skegness
1941 (Banff)—H M King George VI, Sandringham
1942 (Banff)—W J Smith, Leicester
1943 (Banff)—W L Thackray, Malton
1944 (Banff)—A Bush, Tibshelf
1945 (Thurso)—Jepson Bros & Curtis, Peterborough
1946 (Lerwick)—Perry, Morgan & Headman Bros
1947 (Lerwick)—R Preston, Ellistown
1948 (Lerwick)—A Bush, Tibshelf
1949 (Lerwick)—S Powling, Ipswich
1950 (Lerwick)—E Steveson, Mansfield
1951 (Lerwick)—HM King George VI, Sandringham
1952 (Lerwick)—V Divit, Mansfield
1953 (Lerwick)—H Harding, Spalding, Lincolnshire
1954 (Lerwick)—Mrs H A Bridge, Thundersley, Essex
1955 (Lerwick)—A Keeble, Ipswich
1956 (Lerwick)—H A Bridge, Thundersley, Essex
1957 (Lerwick)—Marsh & Bailey, Swanswick, Derbyshire
1958 (Lerwick)—A Smith, Mansfield
1959 (Lerwick)—Mrs H A Bridge, Thundersley, Essex
1960 (Lerwick)—Carter & Son, Thorney
1961 (Lerwick)—E Soames, Manningtree, vel 1219
1962 (Lerwick)—Caston & Son, Brundall, vel 1233 ypm
1963 (Lerwick)—G Breuninger, Radcliffe
1964 (Lerwick)—F H Perkins, Boston
1965 (Lerwick)—J H Lovell, Donington, vel 752
1966 (Thurso)—Springthorpe Bros, Hucknall, vel 1019
1967 (Lerwick)—Waldram & Young, Alfreton
1968 (Lerwick)—F Hudson, Lincoln
1969 (Lerwick)—N Springthorpe, Hucknall, vel 947
1970 (Lerwick)—J & R Brill, Ipswich (731 ypm)
1971 (Lerwick)—Smith & Coupland, Alford, vel 1235
1972 (Lerwick)—J H Lovell, Donnington, vel 840
1973 (Lerwick)—W A Lawson, Ravenstone, vel 868
1974 (Lerwick)—Mr & Mrs C Wright, Bourne, vel 789
1975 (Lerwick)—Wigg Bros, Knodishall, vel 1678
1976 (Lerwick)—R Brunt, Walesby, vel 1085
1977 (Thurso)—Mr & Mrs J Brand, Ely, vel 1118
1978 (Lewick)—Bates, Son & Burley, vel 1989
1979 (Lerwick)—R H Todd, Boston, vel 1398
1980 (Lerwick)—A G Pearce, South Raynham, vel 959
1981 (Lerwick)—Mr & Mrs C Maycock, vel 1935
1982 (Lerwick)—F Parkinson & Son, Spalding, vel 1327
1983 (Lerwick)—D Shore, Market Deeping, vel 1096
1984 (Lerwick)—Mr & Mrs P Barker, Caister, vel 1743
1985 (Lerwick)—J Hammond, Sutton Bridge, vel 1280
1986 (Lerwick)—Mr & Mrs C Adcock, Stapleford, vel 1014
1987 (Lerwick)—Bates, Son & Burley, Langley Mill, vel 1097
1988 (Lerwick)—Mr & Mrs O Payne, King's Lynn, vel 1504
1989 (Lerwick)—J H Lovell, Lincoln, vel 1155
1990 (Lerwick)—H Taylor & Sons, Eastwood, vel 1112
(Thurso)—S Tristram, Borrowash, vel 978
1991 (Lerwick)—J Cotterill, Boughton, vel 1341
(Thurso)—C A Harrison, Norwich, vel 1439

1992 (Lerwick)—G Dobb, Sutton in Ashfield, vel 1278
(Thurso)—Barrett Bros, Norwich, vel 940

SCOTTISH NATIONAL FC

1900 (Skibbereen)—J H Johnstone, Uddingston
1901 (Skibbereen)—J E Taylor, Langside
1902 (Skibbereen)—G Hamilton, Govan
1903 (Bath)—G Hamilton, Govan
1904 (Weymouth)—J Tennant, Bothwell
1905 (Guernsey)—Wyper & Tait, Quarter
1906 (Guernsey)—T Chambers, Motherwell
1907 (Guernsey)—Muir Bros, Newmilns
1908 (Granville)—W Lindsay, Auchengray
1909 (Granville)—Smart Bros, Shieldhill
1910 (Rennes)—W McLean, Dennyloanhead
1911 (Rennes)—J McMeekin, Glasgow
1912 (Rennes)—T Paton, Galston
1913 (Rennes)—A C Christie, Leslie
1914 (Rennes)—Sharp Bros, Newtongrange
1915 and 1916—No Races
1917 (Salisbury)—Dr W Anderson, Armadale
1918 (Weymouth)—Dr W Anderson, Armadale
1919 (Weymouth)—R Birrell, Baillieston
1920 (Rennes)—Dr W Anderson, Armadale
1921 (Rennes)—Watson Bros, Armadale
1922 (Rennes)—W Gardiner, Glasgow
1923 (Rennes)—White & Dickson, Paisley
1924 (Rennes)—J Birrell, Renfrew
1925 (Rennes)—J Brownlie, Carluke
1926 (Rennes)—D McInroy, Dundee
1927 (Rennes)—McGinn Bros, New Cumnock
1928 (Rennes)—F Murdoch, Baillieston
1929 (Rennes)—D Macauley, Dunfermline
1930 (Rennes)—J Robertson, Slamannan
1931 (Rennes)—R Duncanson, Leven
1932 (Rennes)—Swain Bros, Markinch
(Nantes)—Captain D Leslie, Gourock
1933 (Rennes)—J Laidlaw, Broxburn
(Nantes)—J Gillespie, Airdrie
1934 (Rennes)—Frame & Cochrane, Strathaven
(Nantes)—M Thorburn, Lockerbie
1935 (Rennes)—Casey Bros, Strathaven
(Nantes)—Rorison & Sweden, New Cumnock
1936 (Rennes)—J Russell, Quarter
(Nantes)—J Aitchison, Annan
1937 (Rennes)—J Kirkpatrick, Annan
(Nantes)—A Neil, Lugar
1938 (Rennes)—H Park, Armadale
(Nantes)—Smith & Aitken, Braidwood
1939 (Rennes)—Anderson Bros, Shieldhill
(Nantes)—G Francis, Fauldhouse
1940 to 1945—No Races
1946 (Hastings)—A Paterson, Baillieston
1947 (Rennes)—N Campbell, Dunbar
1948 (Guernsey)—W Gray, Chryston
1949 (Charleroi)—G Mitchell, Gartcosh
1950 (Rennes)—G Lupton, Annan
1951 (Rennes)—W J McKay, Annan
(Nantes)—D Steel, Canonbie
1952 (Rennes)—J Marshall, Hamilton
(Nantes)—J Kirkpatrick, Annan
1953 (Rennes)—Montgomery Bros, Catrine
(Nantes)—J R Turner, Carstairs
1954 (Rennes)—D Fowler, Methil, Fife
(Nantes)—Dr H O Martin, Jedburgh
1955 (Rennes)—J Hodgson, Drem
(Nantes)—J Clydesdale, Ormiston
1956 (Rennes)—A Galloway, Leslie, Fife
(Nantes)—J McGillivray, Forth
1957 (Rennes)—G Hay, Bo'ness
(Nantes)—H Scott, Milton, Fife
1958 (Rennes)—Thomson & Kean, Annan
(Nantes)—J McGillivray & Son, Forth
1959 (Rennes)—Irvine & Johnstone, Annan
(Nantes)—J Macpherson, Brechin
1960 (Rennes)—J Wilson, Lanark
(Nantes)—F Trzebniak, Dunbar

1961 (Rennes)—J & F Dora, Arbroath, vel 1414 ypm
(Nantes)—Ferguson Bros & Forsyth, vel 890
1962 (Rennes)—D Angus, Symington, vel 890
(Nantes)—J B Halliday, Brydekirk, vel 1387
1963 (Rennes)—Mr & Mrs Horsburgh, Ecclefechan
(Beauvais)—G Robertson, Cupar
1964 (Rennes)—Shillinglaw & Wylie, Newcastleton
(Nantes)—W O'Neill, Hamilton
(Beauvais)—J Mitchell, Bishopsbriggs
1965 (Rennes)—W Gardiner, Lockerbie, vel 1074
(Nantes)—Beglin Bros, Bo'ness, vel 844
(Beauvais)—Hind & Crombie, Annan, vel 1003
1966 (Rennes)—M Nash & Son, Croy, vel 1086
(Avranches)—T Little, Creca, vel 946
(Nantes)—J Hamilton, Kennoway, vel 920
1967 (Nantes)—J Roach & Son, Broxburn, vel 1146
(Rennes)—R Hope, Eastriggs, vel 1645
(Avranches)—Innes Bros, Gilmerton, vel 1133
1968 (Rennes)—T Harper, Stranraer, vel 1350
(Tours)—J Williams, Gilmerton, vel 617
(Avranches)—P Virtue, Cockburnspath, vel 938
1969 (Rennes)—J Ross, Haddington, vel 1113
(Avranches)—Joe Newcombe, Gordon, vel 1290
(Nantes)—M Jamieson, Annan, vel 1046
1970 (Rennes)—Knox Bros, Dunbar, vel 1344
(Avranches)—D Anderson, Carnwath, vel 969
(Nantes)—J Robertson, Jedburgh, vel 1389
1971 (Rennes)—J D Ellwood, Langholm, vel 1325
(Avranches)—J McClelland, Annan, vel 978
(Nantes)—J T Jamieson & Son, Annan, vel 870
1972 (Rennes)—R & H Kennedy, Irvine, vel 1587
(Nantes)—J Robertson, Jedburgh, vel 1258
(Avranches)—G Sneddon, Dunfermline, vel 1003
1973 (Rennes)—R Rutherford, Coldingham, vel 1310
(Nantes)—W & D Ree, Broughty Ferry, vel 1040
(Avranches)—A Kerr & Son, Danderhall, vel 1048
1974 (Rennes)—E Dishington, Reston, vel 1248
(Nantes)—Mr & Mrs Muirhead, Dalbeattie, vel 835
(Avranches)—J Fairburn, Cockburnspath, vel 1051
1975 (Avranches)—J Watson, Kirkliston, vel 1120
(Rennes)—J Poole, Newmains, vel 983
(Nantes)—T Hopwell, St Boswells, vel 1104
1976 (Avranches)—W Bolton, Cockburnspath, vel 1164
(Rennes)—J Murray & Son, Ecclefechan, vel 1149
(Nantes)—W Halliday, Brydekirk, vel 1069
(Avranches)—W Masson, Inverallochy, vel 1107
(Falaise)—J M Dalgleish, Annan, vel 1468
(Rennes)—T Pringle, Selkirk, vel 1166
1977 (Nantes)—R L Whyte, Annbank, vel 1648
(Avranches)—J Wallance, Ayr, vel 1239
1978 (Falaise)—Shillinglaw & Wylie, Newcastleton,
vel 1023
(Rennes)—George Gray, Cockburnspath, vel 1063
(Nantes)—J Crombie, Annan, vel 1016
(Avranches)—J Fairburn, Cockburnspath, vel 1132
1979 (Falaise)—J McKinnon, Ladybank, vel 963
(Rennes)—Newcombe Bros, Macmerry, vel 955
(Nantes)—Anderson & Tennant, Annan, vel 1137
(Avranches)—L Wood, Duns, vel 1106
1980 (Avranches)—E Newcombe, Macmerry
(Rennes)—M White & Son, Paisley, vel 1092
(Nantes)—R Chapman, Chirnside, vel 900
1981 (Avranches)—S Peaston & Son, South Queensferry,
vel 1001
(Rennes)—D Newcombe, Macmerry, vel 1009
(Nantes)—J T Sanderson, Pathhead, vel 1469
(Sartilly)—Wolf & Thompson, Dunbar, vel 1090
1982 (Sartilly)—M Jamieson, Annan, vel 1112
(Rennes)—Shillinglaw & Wylie, Newcastleton,
vel 1269
(Nantes)—J T Sanderson, Pathhead, vel 970
(Sartilly)—D Rodgerson, Selkirk, vel 1206
1983 (Exeter)—J & W Gray, Cockburnspath, vel 1627
(Dorchester)—W McAuley, Jedburgh, vel 1188
(Dorchester)—W Ferguson & Son, Annan, vel 1339
(Dorchester)—Brown & Gormley, High Valleyfield,
vel 1261

1984 (Sartilly)—Anderson & Tennant, Annan, vel 1228
    (Rennes)—E Newcombe, Macmerry, vel 1180
    (Nantes)—R Bearhope, Coldstream, vel 1291
1985 (Sartilly)—D White & Son, Dumfriesshire, vel 1124
    (Rennes)—J P Jackson, E Lothian, vel 1220
    (Nantes)—Davidson & Little, Cockenzie, vel 1085
1986 (Sartilly 1)—A Forbes, Prestonpans, vel 1161
    (Rennes)—E Callaghan & Son, Fauldhouse, vel 1237
    (Nantes)—M Jamieson, Annan, vel 1056
    (Sartilly 2)—E Hodgson & Sons, Anna, vel 1081
1987 (Sartilly 1)—E Hodgson & Sons, Annan, vel 1177
    (Rennes)—Mr & Mrs I Henderson, Edinburgh,
    vel 1171
    (Nantes)—J Cosgrove & Son, Lesmahagow, vel 1129
    (Sartilly 2)—G McAloney & Son, Coatbridge,
    vel 693
1988 (Sartilly 1)—J Haig, Bilston, vel 1137
    (Rennes)—G Grant & Son, Gretna, vel 988
    (Nantes)—Donaldson & Downie, Rattray, vel 1466
    (Sartilly 2)—H Harper, Dunbar, vel 984
1989 (Sartilly 1)—R Carruthers, Bonnyrigg, vel 1189
    (Rennes)—Gunn & Cherrie, Roslin, vel 1093
    (Niort)—J Duthie, Methil, vel 920
    (Sartilly 2)—R & M Seaton & Son, Annan,
    vel 1412
1990 (Sartilly 1)—P E Heslop, Annan, vel 1449
    (Rennes)—E Hodgson & Sons, Annan, vel 1161
    (Niort)—D N Dall, Ladybank, vel 792
    (Sartilly 2)—E Hodgson & Sons, Annan, vel 1210
1991 (Sartilly 1)—E Hodgson & Sons, Annan, vel 1104
    (Rennes)—Glendinning & Edwards, Gretna, vel 1095
    (Niort)—J S Irving, Brydekirk, vel 1529
    (Sartilly 2)—Lytollis & Stevenson, Gretna, vel 1075
1992 (Sartilly 1)—J Murphy & Son, Kirkcaldy, vel 1237
    (Rennes)—J Honeyman & Son, Kennoway, vel 1163
    (Nantes)—Mr & Mrs J Graham, Ecclefechan, vel 854
    (Sartilly 2)—G Campbell, Annan, vel 1422

BRITISH INTERNATIONAL CHAMPIONSHIP CLUB

1978 (Pau)—C H Wells, Ramsgate, vel 861
    (Barcelona)—C T Galyer, Wallington, vel 798
1979 (Pau)—Fright & Johnsons, Minster, vel 898
    (Barcelona)—Fear Bros, Clandown, vel 798
1980 (Pau)—J Shepherd, Chichester, vel 755
    (Barcelona)—C H Wells, Ramsgate, vel 762
1981 (Pau)—A E Shepherd, Chichester, vel 846
    (Barcelona)—C H Wells, Ramsgate, vel 896
1982 (Pau)—Cecil Bros, Ramsgate, vel 1039
    (Pau)—M J Humphrey, Forest Row, vel 798
1983 (Scilly Isles)—S G Biss, Brundall, vel 992
1984 (Pau)—J & R Wills, Feltham, vel 800
    (Barcelona)—D Bartley, Ashwell, vel 849
1985 (Pau)—G Hunt, Westmarsh, vel 993
    (Barcelona)—Mr & Mrs Newman, vel 818
1986 (Pau)—G Hunt & Son, Westmarsh, vel 876
    (Barcelona)—Shepherd & Buckley, Chichester,
    vel 755
1987 (Pau)—G M Young, Timsbury, vel 801
    (Barcelona)—J Lane, Bromley, vel 694
1988 (Pau)—T Dodd, Taunton, vel 833
    (Barcelona)—G Hunt & Son, Westmarsh, vel 1076
1989 (Pau)—W Ross, All Stretton, vel 856
    (Barcelona)—T Twyman & Son, Harrow, vel 555
1990 (Pau)—G Hunt & Son, Kent, vel 916
    (Barcelona)—E Deacon, Hants, vel 671
1991 (Pau)—B & R Long & Cox, Stowmarket, vel 972
    (Barcelona)—D A Delea, Rainham, vel 908
1992 (Pau)—R J Clarke, Clyffe Pypard, vel 855
    (Barcelona)—G Hunt & Son, Westmarsh, vel 850

WELSH GRAND NATIONAL FC

1944 (Fraserburgh)—F G Morgan
1945 (Fraserburgh)—J Tucker
1946 (Thurso)—Owen Bros & Morris
1947 (Thurso)—I Gardner
1948 (Thurso)—J Reynolds

    (Lerwick)—T Clements
1949 (Thurso)—Morgan Bros & Absalom
    (Lerwick)—Parfitt & Son
1950 (Thurso)—Channing & Son
    (Lerwick)—Atwell Bros
1951 (Thurso)—Atwell Bros
    (Lerwick)—Atwell Bros
1952 (Thurso)—Daniels Bros
    (Lerwick)—Atwell Bros
1953 (Thurso)—Morgan & Jones, Newport
    (Lerwick)—T Hibbard, Cwmcarn
1954 (Thurso)—A & L Morgan, Caerphilly
    (Lerwick)—J L Griffiths, Llantarnam
1955 (Thurso)—M Probert, Pontnewynydd
    (Lerwick)—Atwell Bros, Newport
1956 (Thurso)—D Morris, Neath
    (Lerwick)—S J Jones, Barry
1957 (Thurso)—W H George, Pontypridd
    (Lerwick)—Lewis & Evans, Swansea
1958 (Thurso)—W Jaworski, Tynant
    (Lerwick)—Buffett & Rees, Pontypridd
1959 (Thurso)—E Rees, Tylorstown
    (Lerwick)—G Cecil, Pontypridd
1960 (Thurso)—R Hendrickson, Cardiff
    (Lerwick)—Channing & Son, Newport
1961 (Thurso)—Jones Bros, Cwmgwrach, vel 1003
    (Lerwick)—H Knapp, Crumlin, vel 998
1962 (Lerwick)—J Lloyd, Treharris, vel 864
    (Thurso)—Sims, Pye & Evans, Caerau
1963 (Lerwick)—J Davies, Abercynon
    (Thurso)—Walton & Son, Llanbradach
1964 (Lerwick)—K Ball, New Tredegar
    (Thurso)—Bryn Davis, Dowlais
1965 (Lerwick)—F D Alcock, Abercynon, vel 643
    (Thurso)—Edwards Bros, Aberfan, vel 1575
1966 (Elgin)—Harry Hopps, Barry, vel 932
    (Thurso)—Bale Bros & Griffiths, vel 1084
1967 (Lerwick)—Emmet Bros, Pontypool, vel 1468.56
    (Thurso)—R E Jones, Newport, vel 989
1968 (Lerwick)—R E Jones, Newport, vel 1170
    (Thurso)—Woodbridge Bros, Rhydefelin, vel 1191
1969 (Thurso)—I Marshall, Newport, vel 1193
    (Lerwick)—Yews & Thomas, Trehafod, vel 819
1970 (Thurso)—D Gardener & Son, Tredegar
    (Lerwick)—Phillips Bros, Cwmparc
1971 (Thurso)—Davies, Mould & Williams, vel 943
    (Lerwick)—Fear Bros, Pontypridd, vel 431
1972 (Thurso)—Thomas & Son, Merthyr, vel 1049
    (Lerwick)—Sheehy & Son, Tredegar, vel 795
1973 (Thurso)—Radford & Son, Abertysswg, vel 1395
    (Lerwick)—Bowen Bros, Porthcawl, vel 1119
1974 (Thurso)—Jones & Jarman, Trehafod, vel 1359
    (Lerwick)—R Davies, Sandfields, vel 1455
1975 (Thurso)—Spiller, Iles & Billen, Trealaw, vel 1195
    (Lerwick)—Williams & Son, Hirwaun, vel 1420
1976 (Lerwick)—Turton & Son, Ynyshir, vel 902
    (Thurso)—M Simmonds, Llangeinor, vel 1038
1977 (Lerwick)—H O Jones, Gilfach Goch, vel 1465
    (Thurso)—Williams Bros, Tredegar, vel 814
1978 (Lerwick)—F Hulbert, Talgarth, vel 1783
    (Thurso)—M Culverhouse, Baglan, vel 1065
1979 (Lerwick)—B Lewis, son & daughter, Pontlottyn
    (Thurso)—I Jones, Pill, vel 1206
1980 (Lerwick)—M Culverhouse, Baglan, vel 789
    (Thurso)—B & D Jones & Son, Blaenavon
1981 (Lerwick)—Bowen Bros, Porthcawl, vel 1367
    (Thurso)—G Hackling, Islwyn, 1153
1982 (Lerwick)—G R Thomas, Merthyr Tydfil, vel 747
    (Thurso)—A Reed, Blackwood, vel 969
1983 (Lerwick)—Robertson & Yeoman, vel 1122
    (Thurso)—Adams Bros, Blaenclydach, vel 1259
1984 (Lerwick)—A Keevil, Caerphilly, vel 1171
    (Thurso)—Thomas & Son, Merthyr, vel 1084
1985 (Thurso)—Bale Bros & Griffiths, Pill, Newport
    (Lerwick)—G H Williams & Son, vel 819.539
1986 (Lerwick)—Preece Bros, Cwmtillery, vel 1521
    (Thurso)—K Darlington, Barry, vel 1188

1987 (Lerwick)—Jude & Prosser, Cwmbran, vel 1540
    (Thurso)—R Edwards & Son, Port Talbot, vel 943
1988 (Lerwick)—G Daniels, Ystalyfera, vel 868
    (Thurso)—Sterio Bros, Cardiff, vel 1355
1989 (Lerwick)—C Howells & Son, Risca, vel 928
    (Thurso)—Attwell Bros, Cwmbran, vel 1173
1990 (Lerwick)—Jones & Winstone, Pontypool, vel 760
    (Thurso)—J Smale, Abertillery, vel 1020
1991 (Lerwick)—C Howells & son, Risca
    (Thurso)—Fishlock & Harding, Cardiff
1992 (Lerwick)—Birds returned
    (Thurso)—D Rumph & Son, Maesteg

BRITISH BARCELONA CLUB

1965 (Barcelona)—Mr & Mrs E A Rivett, vel 399
1966 (Barcelona)—Mr & Mrs S P Nash, Southampton, vel 726
1967 (Palamos)—A G M Stevens, Swindon, vel 849
1968 (Palamos)—L S Bunn, Folkestone, vel 886
1969 (Palamos)—Mr & Mrs A Hustler, Poole, vel 785
1970 (Palamos)—W J Challen, Cookham Dean, vel 898
1971 (Palamos)—J & J G Paley, Silsden, vel 788
1972 (Palamos)—C Medway, Southampton, vel 637
1973 (Palamos)—R Churchill, Weymouth, vel 627
1974 (Palamos)—T W Hodges, Taunton, vel 1007
1975 (Palamos)—T Dodd, Taunton, vel 860
1976 (Palamos)—T Perrett, Mere, vel 741
1977 (Palamos)—W Bradford, Sutton, vel 843
1978 (Palamos)—R Dowden, Portsmouth, vel 863
1979 (Palamos)—Fear Bros, Clandown, vel 804
1980 (Palamos)—C Medway, Southampton, vel 814
1981 (Palamos)—A G Mulholland & Son, Bristol, vel 856
1982 (Perpignan)—G Hunt & Son, Westmarsh, vel 1170
1983 (Thurso)—W Elgey, Market Wheaton, vel 923
1984 (Palamos)—K Thorpe, Colchester, vel 791
1985 (Palamos)—L Peart & Son, Bradford, vel 545
1986 (Palamos)—E J Vowles, Glastonbury, vel 1048
1987 (Palamos)—G Byrne, Sons & Dtr, Kingston, vel 501
1988 (Palamos)—W Stephenson, Shaftesbury, vel 880
1989 (Palamos)—A Hustler, Poole, vel 882
1990 (Palamos)—Mr & Mrs D R Davis, Landford, vel 859
1991 (Palamos)—M White, Hamworthy, vel 1080
1992 (Perpignan)—D Coulter, Worthing

IRISH NATIONAL FC

1932 Robert Hawthorne, Wolfhill, Belfast
1933 E A Robinson, Shaws Bridge, Belfast
1934 J M K McGugan, Muckamore, Co Antrim
1935 F W Pelan, Lambeg, Co Antrim
1936 John McCormick, Lurgan
1937 E A Robinson, Lurgan
1938 McBride & Brown, Belfast
1939 F R Best, Lurgan
1943 Joseph Smith, Belfast
1946 (Guernsey)—S Mayberry, Antrim
1947 (Le Bourget)—Mrs E Frazer, Belfast
1948 (Rennes)—S Robinson, Belfast
1949 (Rennes)—McDowell Bros, Dromore
1950 (Rennes)—D Cochrane & Son, Banbridge
1951 (Rennes)—F Phillips, Belfast
1953 (Redon)—McDowell & Son, Ards
1954 (Redon)—Thompson Bros, Gilford
1955 (Redon)—Parkinson Bros, Lisburn
1956 (Rennes)—N Petts & Partners, Comber
1957 (Redon)—H Spratt, Belfast
1958 (Redon)—Chambers & Smith, Seapatrick
1959 (Redon)—G McCartney Bros, Moira
1960 (Les Sables)—L Burns, Dromore
1961 (Les Sables)—A Blair, Carrickfergus, vel 820
1962 (Les Sables)—S Banks Snr, Moira, vel 1116.8
1963 (Les Sables)—A McDonnell, Coleraine
1964 (Nantes)—Gowdy Bros, Muckamore, Co Antrim
1965 (Nantes)—McKnight & Son, Banbridge
1966 (Nantes)—Doran Bros, Gilford
1967 (Nantes)—R Dunlop, Castlereagh
1968 (Nantes)—W Erwin, Ballymena
1969 (Nantes)—N Corry, Dunmurry

1970 (Nantes)—W McCluggage & Son
1971 (Nantes)—S Bell, Dundonald
1972 (Nantes)—T Cairns, Monkstown
1973 (Nantes)—McDowell & Son, Newtonards, vel 518
1974 (Nantes)—A Simpson & Son, Randalstown, vel 711
1975 (Rheims)—S Tupping, Donacloney, vel 735
1976 (Rennes)—F Adams, Belfast, vel 898
1977 (Rennes)—G Lyons, Lisburn, vel 1526
1978 (Rennes)—Connor Bros, Dundrum, vel 982
1979 (Rennes)—T Spears, Dublin, vel 1125
1980 (Les Sables)—G Douglas, Lurgan, vel 954
1981 (Les Sables)—E Reid & Son, Dublin, vel 1111
1982 (Les Sables)—Joe Doheney, Dublin, vel 914
1984 (Thurso)—Connor Bros, Dundrum, vel 1005
    (Lerwick)—Russell & Moffat, Banbridge, vel 641
1985 (Thurso)—Mrs A Cheevers, Crossgar, vel 1143
    (Lerwick)—J Crossan, Downpatrick, vel 605
1986 (Jersey)—A Darragh, Cullybackey, vel 891
1987 (Jersey)—Campbell & Francey, Harryville, vel 1328
1988 (Rennes)—A McDonald, Portadown, vel 1333
1989 (Rennes)—R Carson & Son, Banbridge, vel 1087
1990 (Rennes)—G A Willis, Portadown, vel 778
1991 (Rennes)—B Evans, Dundalk, vel 963
    (Sartilly)—Smyth Bros, Ballymena, vel 700
1992 (Rennes)—J Cullen, Dublin
    (Sartilly)—R Clements, Ballymena, vel 558

UP NORTH COMBINE

1921 (Mons)—Brass & Bruce, Easington Lane
    (Troyes)—Donaldson, Seaham
1922 (Mons)—Hardy, Hetton Downs
    (Troyes)—Brass & Bruce, Easington Lane
    (Chimay)—Atkinson, Houghton
    (Nevers)—Robson, Hirst
1923 (Mons)—J M Neasham, Dawdon
1924 (Troyes)—Jones, Easington Colliery
    (Chimay)—Phillips, South Shields
1925 (Melun)—Smith & Son, South Shields
    (Nevers)—W Johnson, Gilesgate Moor
1926 (Melun)—Allen Bros, Gosforth
    (Nevers)—J Ellis, Gosforth
1927 (Melun)—Dowson, Sunniside
    (Nevers)—Peat Bros, Bishop Auckland
1928 (Melun 1)—Lothian, Embleton & H, Newbiggin
    (Melun 2)—A Wilson, Hirst
    (Nevers)—Davidson Bros, Arthur's Hill
1929 (Melun)—McKee & Son, Easington Colliery
    (Nevers)—Dodds & Stewart, Morpeth
1930 (Nevers)—Collins Bros, Wrekenton
    (Le Puy)—Lothian & Embleton, Newbiggin
1931 (Nevers)—Dryden & Cochrane, South Shields
    (Le Puy)—J O'Neill, Hexham
1932 (Nevers)—W Baston, Alnwick
    (Le Puy)—Hickman & Simpson, Forest Hall
1933 (Melun)—Potts & Cairns, Easington Colliery
    (Nevers)—Johnson & Brown, Sunderland
1934 (Nevers)—Cobbledick & Wallace, Ashington
    (Le Puy)—Jones, Shotton
1935 (Melun 1)—Gray Bros, Hirst
    (Melun 2)—Carter & Clark, Sunderland
    (Nevers)—Soulsby & Son, Chester-le-Street
1936 (Melun 1)—R Allan, Amble
    (Melun 2)—Ellison Bros, Easington
    (Nevers)—A Hallimond, Dawdon
1937 (Melun)—Terry Bros, Blackhall
    (Nevers)—J Ward, New Seaham
1938 (Melun)—Rowe & Rowe, Haswell
    (Nevers)—Hayton & Son, Houghton
1939 (Melun)—T Gibson, Amble
    (Nevers)—W Heyden, Dawdon
1940 to 1946—No Races
1947 (Lille)—W Summers, Broomhill
    (Le Bourget)—T Pilmour, Hartlepool
1948 (Guernsey 1)—Hird & Graham, Framwellgate Moor
    (Guernsey 2)—Robson & Son, Easington
    (Guernsey 3)—M Ward, Blackhall
1949 (Brussels 1)—R Morris, North Shields

(Brussels 2)—Hewitson & Son, North Seaton
(Luxembourg)—A Bilverstone, North Shields
1950 (Brussels 1)—Bland & Son, Ashington
(Brussels 2)—J Shemmings, Sunderland
(Luxembourg)—Short, Dunn & Reynolds, Horden
1951 (Brussels 1)—Kilpatrick, Whitley Bay
(Brussels 2)—Brannan & Partners, Sunderland
(Luxembourg)—J Peach & Son, Sunderland
1952 (Brussels 1)—Forbes Bros & Ramsey, Horden
(Brussels 2)—Johnson & Shillito, Ryhope
(Luxembourg)—J Hall, Skinningrove
1953 (Brussels 1)—Knox & Son, Broomhill
(Brussels 2)—Newby & Robinson, New Seaham
(Luxembourg)—R Nichols, Dudley
1954 (Le Bourget)—Tinker, Son & Piggford, Blackhall
(Bourges)—H James, Jarrow
1955 (Cormeilles)—Stewart & Barnes, Castletown
(Bourges)—Clark Bros, Pegswood
1956 (Cormeilles)—Green Bros & Fenwick, Dawdon
(Bourges)—Heydon Bros, Dawdon
1957 (Cormeilles)—G Foster, Tudhoe
(Bourges)—Robinson & Partners, N Biddick
1958 (Bourges)—Howells & Partner, Ferry Hill
(Angouleme)—Duff Bros, Easington
1959 (Bourges)—Laverick, Dawdon W.
(Angouleme)—Pringle & Deignam, Jarrow
1960 (Cormeilles)—J Hall, Skinningrove
(Bourges)—Wood Bros, Skinningrove
1961 (Cormeilles)—Hicks, Sherwood & Hansell, Staithes
(Bourges)—Kilner & Donkin, Horden
1962 (Bourges)—Douglas & Little, Chevington Drift
(Cormeilles)—J & W Douglas, Holy Island
1963 (Beauvais 1)—W Fraser, Walker
(Bourges)—D Smith, Great Ayton
(Beauvais 2)—T C Anderson, Newcastle
1964 (Beauvais)—A Proctor, Heworth
(Angouleme)—Kilner & Donkin, Horden EE
1965 (Bourges)—Porritt Bros, Staithes, vel 1091
(Beauvais)—F Cummings, Dawdon, vel 1135
1966 (Evereux)—W Thompson & Son, Alnwick, vel 1399
(Bourges)—J Hoggard, Loftus, vel 1359
1967 (Angeville)—Chapman Bros, Easington Colliery, vel 1324
(Bourges)—Marshall, Redcar, vel 1109
1968 (Beauvais)—Thompson, Seaton Sluice, vel 1055
(Bourges)—Orley & Son, Hartlepool, vel 1502
1969 (Beauvais)—Burn Bros, South Hetton, vel 1257
(Bourges)—Burton Bros, Hartlepool, vel 1029
1970 (Beauvais 1)—Porter Bros, Washington, vel 1194
(Beauvais 2)—Mr & Mrs Baker, vel 1091
(Bourges)—Tinkler, Son & Piggford, vel 1234
1971 (Beauvais 1)—Wood, Miller & Lauderdale, vek 1394
(Beauvais 2)—Crackett Bros & Donaldson, vel 1158
(Bourges)—Wood Bros, Blythe Boro (1349 ypm)
1972 (Beauvais 1)—Leach & Brown, Boosbeck, vel 1442
(Beauvais 2)—J Stevenson, Berwick, vel 1131
(Bourges)—K Welford, Lingdale, vel 1063
1973 (Beauvais)—Scott Bros, Blyth, vel 1696
(Bourges)—McGowan & Kirkbride, vel 1401
(Beauvais 2)—A Wilkie, Stobswood, vel 1103
1974 (Beauvais 1)—Pentz & Johnston, Wallsend, vel 1209
(Bourges)—Mrs J Barker, Boosbeck, vel 994
(Beauvais 2)—Laws & Maw, Cornhill, vel 1222
1975 (Beauvais 1)—Huntingdon & Son, Shotton, vel 1145
(Bourges)—Shell, Son & Hedley, Alnmouth, vel 1017
(Beauvais 2)—E Rowe & Son, Skinningrove, vel 1220
1976 (Beauvais 1)—E Rowe & Son, Skinningrove, vel 1270

(Bourges)—Borrowdale & Son, Usworth, vel 964
(Beauvais 2)—Rutherford & Hart, Bedlington, vel 1201
1977 (Beauvais 1)—Thompson Bros, Lingdale, vel 1598
(Bourges)—Richardson & Son, Holywell, vel 1083
(Beauvais 2)—Anderson & Son, Wingate, vel 1028
1978 (Melun 1)—D McCabe, New Heaton, vel 1044
(Bourges)—Moorhead & Son, Owton Moor, vel 1231
(Melun 2)—A J Gill, Guisborough, vel 1096
1979 (Clermont)—Dixon & Wood, Shiremoor, vel 1272
(Bourges)—Salloway & Son, Hartlepool, vel 1372
(Melun)—Kelly & Son, Hartlepool, vel 1118
1980 (Beauvais)—Soderland & Bradley, West Park, vel 1269
(Bourges)—Miller & Son, Ashington, vel 1181
(Melun)—J Nicholson, Whiteleas, vel 1098
1981 (Beauvais)—G Hoggarth, Skinningrove, vel 1538
(Bourges)—Miller Bros & Lauderdale, Horden, vel 1172
(Melun)—Basley & Roberts, Horden, vel 1249
1982 (Beauvais)—Arkle & Scott, Bensham, vel 1078
(Bourges)—Watson & Clayton, Lingdale, vel 1110
(Melun)—Watson & Clayton, Lingdale, vel 1303
1983 (Eastbourne 1)—D Wilkinson, Newbiggin, vel 1396
(Weymouth)—Hagan & Son, Southbank, vel 1252
(Eastbourne 2)—Freeman Bros, Wreckenton, vel 1293
1984 (Bourges)—J Nicholson, Whiteleas, vel 895
(Lillers)—C Wardell, Skelton Green, vel 1129
(Beauvais)—J Hall & Son, Seaton Sluice, vel 1268
(Melun)—Hume & Son, Loftus, vel 1411
1985 (Beauvais)—T McCluskey & Son, Southwick, vel 1187
(Bourges)—Carr & Weatherstone, Tynemouth, vel 1109
(Melun)—Davidson & Adams
1986 (Beauvais)—Thurlow, McCluskey & Son, Southwick, vel 1187
(Bourges)—Carr & Weatherstone, Tynemouth, vcl 1109
(Melun)—Brewster & Speckman, Wingate, vel 1123
1986 (Beauvais)—W Bell & Son, Newbiggin, vel 1157
(Bourges)—W Gates & Son, Boldon, vel 1050
(Melun)—Hindhaugh, Son & Donaldson, Celtic, vel 1214
1987 (Beauvais)—Dawson Bros & Stephenson, Owton Manor, vel 1341
(Bourges)—C Copeland, Murton Cornwall, vel 1050
(Melun)—Arkle & Scott, Bensham, vel 1267
1988 (Abbeville)—Kay, Morgan & Son, Meadowell, vel 1290
(Bourges)—F & J Gray, Newbiggin, vel 814
(Melun)—Dodds & Hubbert, Stranton HS, vel 1278
1989 (Beauvais)—T Laskey & Son, Wrekenton, vel 1594
(Bourges)—Lynch & Hintchcliffe, Jolly Farmers, vel 1017
(Clermont)—H Hatfield & Son, Loftus HS, vel 1133
1990 (Clermont)—Coulson & Rose Bros, Castletown
(Bourges)—Mr & Mrs R Seaton & Son, Guisborough
(Provins)—Arkle & Scott, Bensham
1991 (Clermont)—G Wandless, Hetton, vel 1469
(Bourges)—A Cole, Guisborough, vel 1189
(Lillers)—Ali Bros, South Shields, vel 1263
1992 (Beauvais)—Kain & Foster, Bensham, vel 1305
(Provins)—Lilley & Son, Hartlepool, vel 1146
(Bourges)—J Anderson & Sons, Wingate, vel 1162

LIST OF FEDERATIONS

Information Concerning Federations Affiliated to the FCI
(as supplied by the FCI)

Czechoslovakia — Cesky Svaz Chovatelu, Slavibor Petrzilka, Sekretär des ZA CSHC, Maskova 3, 18253 Praha 8 — Kobylisy

Hong-Kong — The Hong-Kong Racing Pigeon Association, no 9 A, Lion Rock Road 4th/F, Kowloon City, Kowloon, Hong-Kong. Phone: Tel: 716.26.85

Hong-Kong — Hong-Kong Region Unit of AFPA, PO Box 2592. **President:** Simon Lee Shu Dong

Malta — **President:** Alfred Naudi, 'De Naudi', Fleur de Lys Road, Bikara — Malte 441766

URSS/VOGS — **President:** Alexandre Kareline, Moscou 121165, 28 rue Staidents Reskaya, bâtiment 2, App 186

USA (IF AHPF) — **President:** Matthew Reilly, 353 Doherty Avenue, Elmont, LI, NY 1003

South Africa — South African Homing Union, 11 Quentin Road, Robertsham, 2091 Johannesburg. Phone: 011 680. **President: S F du Toit**, Eufees Road 29, Bayswater, Bloemfontein 9301. Phone: 05131-1432

Germany (West)* — Verband Deutscher Brieftaubenliebhaber, Schönleinstrasse 43, 4300 Essen 1. Phone: 0201-778071. Fax: 201/77.38.11. **President: Prof Dr Med Jozef Kohaus**, Hülsbergstrasse 23, 4630 Bochum. Phone: 00/49234-791045

Germany (East) — Sektion Sporttauben der DDR, Wichertstrasse 10, Berlin 1071. Phone: 44.818.13. **Président: Ralf Geier**, Eichlörnchenweg 3, Erkner 1250. Phone: 650.67.25
*(Revision pending)

Austria — Verband Osterreichischer, Brieftaubenzuchter-Vereine, Hohe Wandstrasse 1, 2630 Pottschach. Phone: 02622-22147. **President: Ing Lutz Primes**, Waldgasse 788, 7201 Neudorfl ad Leitha. Phone: 02622-77 59 25

Australia — Australian Pigeon Fanciers Association, Dalton Street 36, Towradge, New South Wales 8518. Phone: 6142-847290. **Président: John Swan**, International correspondent: Gen Manager: Geraldine Henshaw, 22 Vernon Street, Greystnes, NSW 2145. Phone: 02.631 6929

Belgium — Royale Federation Colombophile Belge, Rue de Livourne 39, 1050 Bruxelles. Phone: 2/537.62.11. Fax: 2/538.57.21. **President: Marcel van den Driessche**, Van Langenhovestraat 19, 9200 Dendermonde. Phone: 052/21.37.52

Bulgaria — Union Cooperative Centrale, Union Des Cuniculiculteurs Des Colombophiles, and Amis de la Nature, Sofia 1000, 99 rue G S Rakovski. Phone: 8441. **President: Ivan Zachariev.** Phone: 884074

Canada — Canadian Racing Pigeon Union, c/o Dorothy Deveau, 107-246 Blakie Road, London, Ontario N6L 1E4. Phone: 519-652-5704. **President: Ken Hutchison**, 554 Erie Str, Strafford, Ontario, N5A 2N6. Phone: 519-273-2458

Korea — Korean Racing Pigeon Association, 12-33 Wonheung-Ri, Wondang-Eup, Koyang-Kun, Kyungxi-Do, Korea. Phone: 0344-64-3557. **Président: Dr Yang Hee Lee.**

Cuba — Federacion Colombofila De Cuba, Escobar 104 e/Animas y Lagunas, Le Habana 10200 — Ciudad de la Habana. Phone: 70-3942/7104 79-2472. **Président: Guillermo Garcia**, 4th/514 e/5 TD and 7th, Sta Cruz del Norte, La Habana.

Denmark — De Danske Brevdueforeninger, Riddersborg 9, 4900 Nakskov DK. Phone: 53-92.14.42. **President: Erling Vestergaard**, PO Box 69, 9800 Hjorring DK. Phone: 08-92.05.00

Egypt — Egyptian Racing Pigeon Federation, 8 Toschtomor Street, from Talaab Harb Street, Cairo. **President: Dr Samy Ismail**, 1 Amin El Samsy Street, El-Hegaz, Heliopolis — Cairo.

Spain — Real Federacion Colombofila Espanola, Eloy Gonzalo 34, Piso 7º Izda, 28010 Madrid. Phone: 448.88.42. **Président: D Carlos Márquez Prats**, Pau Casals 5-3º, 08021 Barcelona. Phone: 866.08.55

USA — IF of American Homing Pigeon Fanciers Inc, Secr/Trés: Mme Marie Rotondo, 107 Jefferson Street, Belmont Hills, Pa 19004. Phone: 215.664.0266. **President: Nicolas Schiavone**, 5207 Linden Street, Philadelphia, Pa 19114. Phone: 215.637.7766

USA — The American Racing Pigeon Union, Secr/Trés: Russ Burns, PO Box 2713, South Hamilton, Ma 01982. Phone: 508.927.3631. **President: Charles E Weaver**, 624, Kumukahi Place, Honolulu, HI 96825. Phone: 808.395.2566. Bur: 808.841.4100

France — Union Des Federations Regionales Des, Associations Colombophiles De France, Boulevard Carnot 54, 59042 Lille Cédex. Phone: 20.06.82.87. **President: Roger Mezieres**, Rue Ernest Dujardin 21, 59290 Wasquehal. Phone: 20.72.04.67

Luxembourg — Federation Colombophile Luxembourgeoise. **President: Désiré Nick**, 223 Route de Soleuvre, 4670 Differadange/Obercorn. Phone: 585.229

Great Britain — Royal Pigeon Racing Association, The Reddings, Nr Cheltenham, Gloucestershire, GL51 6RN. Phone: 0452 713529. Fax: 45285-7119. **President: J James**, Herbert Lodge, Drybrook, Gloucester GL19 7AD. Phone: 0594 542798

Great Britain — North of England Homing Union, 58 Ennerdale Road, Walker Dene, Newcastle-upon-Tyne NE6 4DG. Phone: 091-262 5440. **President: J E Brown**, 29 Biddick Villas, Washington, Tyne & Wear

Hungary — Magyar Postagalambsport Szovetseg, 1076 Budapest, Verseny Utca 16. Phone: 36-11-424.522. **President: Attila**, 2600 VAC, Cserje u 17.

Italy — Federazione Colombofila Italiana, Via Lustrini 1, 42100 Reggio Emilia. Phone: 0522.25671. Fax: 721-31097. **President: AVV Francesco Paci**, Via Ardizi 14, 61100 Pesar. Phone: 0721.67884

Japan — Japan Racing Pigeon Association, 17-11 Uéno Park, Taito-Ku, Tokyo 110. Phone: 03.822.4231. Fax: 03-822.4234. **President: Ryoichi Sasakawa**, Senpaku Shinko Bldg, L-15-16 Toranomon, Minato-Ku, Tokyo. Phone: 03.502.2371

Malta — Federation of Pigeon Racing Clubs of Malta, Railway Ave, Hamrum. Phone: 230111. **Président: Ernest Rossi**, Villa Andrea, San Andrea Estate, San Gwann — Malta. Phone: 00356-335033

Mexico — Federacion Mexicana de Columbofilia AC, Secr Apartado Postal 1004-H, San Luis Potosi, SLP — Mexico. **Président: Miguel Azcona**, Enrico Martinez 255, 78240 San Luis Potosi — Mexico. Phone: 481-364-28

Norway — Norges Brevdueforbund, V/per Kristian Hansen, Postboks 916, Kraakeroy — 1600 Fredrikstad. Phone: 093-40003. **Président: Arvid Myreng**, Brekke, 3100 — Tonsberg. Phone: 033-30869

Holland — Nederlandse Postduivenhouders Organisatie, Landjuweel 38, 3905 PH Veenendaal. Phone: 838.52.33.91. Fax: 8385-23712. **President: G Pankras**, Clauslaan 20, 1735 HA 't Veld.

Poland — Polski Zwiazek Hodocow Golebi, Pocztowych Zarzad Glowny, 41-501 Chorzow, Wpkiw — Skrytka Pocztowa 62. Phone: 415.984. **President: Jozef Paliczka**, Ul Krzyzowa 8a/22, 40-111 Katowice. Phone: 502.995

Portugal — Federacao Portuguesa de Columbofilia, Rua Padre Estevao Cabral 79 — Salas 214/215, 3000 Coimbra. Phone: 039.36763. **Président: Gaspar Bendito Vila Nova**, Rua Nicolau Chanterene 340 S/C Dta, 3000 Coimbra. Phone: 039.22303

China (Taiwan) — China Racing Pigeon Association, Room No 5, 12th Floor No 112, Chu Shan North Road, Sect 2, Taipei 10449 — Taiwan. Phone: 02581.1322. **Président: Hsin-Yi Chen**

Romania — Uniunea Federatiilor Columbofile din Romania, Strada Fagaras 14, Sector 1, 77108 Bucarest. Phone: 50.44.44. **Président: Ioan Carmazan**, Bld Republicii 77, Ap 1, Sect 2, Bucarest. Phone: 166.328

Sweden — Svenska Brevdueforbundet, Trädgärdsvägen 20, 23700 Bjarred. Phone: 046.294.834, Secrétaire: Agne Nilsson. **President: Viggo Elofsson**, Gringelstadsvägen 110, 29191 Kristianstad. Phone: 044.235074

Switzerland — Association Centrale Des Societes, Colombophiles Suisses, Mürenbergstrasse 1, 4416 Bubendorf. Phone: 061.931.29.70. **President: Ulrich Frei-Zulauf**. Service pigeons égarés: Walter Hofer, Gässli 46, 3472 Wynigen. Phone: 034.55.13.87

Czechoslovakia — Cesky Svaz Chovatelu, Maskova 3, 18253 Praha — Kobylisy. Phone: 84.57.459. **President: Dr Ant Cernosek**, Tokudova 325, 10900 Praha. Phone: 84.67.59

Thailand — Thai Homing Pigeon Federation, 411/2 Nang Linchee Road, Chongnonthree, Yannawa District, Bangkok Metropolis. **President: Paku Tangghaisak**, World Trading Company, 47/49 Mares Road, Bangkok

Trinidad — National Racing Pigeon Commission of Trinidad and Tobago, 3 Scott Bushe Street/Port of Spain, Trinidad — West Indies. Phone: 809.625.9813. **President: Ernest Ferreira**, Bleu Basin Gardens, Diego Martin. Phone: 632.1340

Yugoslavia — Savez Sportskih Klubova Odgajivaca, Golubova Pismonosa SFRJ, 11000 Beograd, Knez Danilova 4. Phone: 05453.390. Secr.Tres: Petrovic. Phone: 011.788.707. **Président: Ivan Vizin**, 2400 Subotica, Krsmanovica 1. Phone: 024.22.717

THE OLYMPIAD SINCE 1948

1949 (Lille)
1. France
2. Holland
3. Scotland
4. Belgium
5. Luxembourg

1951 (Modena)
1. Holland
2. France
3. Belgium
4. Luxembourg
5. Italy

1953 (Copenhagen)
1955 (Barcelona)
1. Holland
2. France
3. Belgium
4. Italy
5. Spain

(1957) (Amsterdam) First year of the official classification Standard Class:
1. Czechoslovakia
2. Holland
3. Belgium
4. Italy
5. Sweden

1959 (Lisbon) The introduction of the Sporting Class as well as the Standard Class:

Standard Class
1. Czechoslovakia
2. Holland
3. Belgium
4. Germany
5. Hungary

Sporting Class
1. Czechoslovakia
2. Germany
3. Belgium
4. Holland
5. Italy

1961 (Essen)
1. Belgium
2. Germany
3. Holland
4. France
5. Poland
6. Britain*

1. Belgium
2. Yugoslavia
3. Germany

*The British team comprised of birds from all over Britain which were nominated and selected through *The Racing Pigeon*.

1963 (Ostend)
Standard Class with 15 countries competing:
1. Czechoslovakia
2. Belgium
3. Britain
4. France
5. Holland.

Sporting Class with seven countries competing:
1. Germany
2. Holland

At this show Trevor Parker's cock Olympic Vision won the Standard Class for cocks with an equal number of points to Mr Chitil, winner of the hens, for Czechoslovakia. It was, therefore, regarded as joint Best in Show.
    The following nine positions were won in the first 25 cocks and first 25 hens in the Standard Class:

1st Cocks—Trevor Parker
3rd Hens—E Griffiths
4th Hens—Trevor Parker
9th Cocks—E Griffiths
10th Hens—Trevor Parker

11th Cocks—L Armitage & Sons
12th Hens—L Armitage & Sons
18th Holmes Bros
20th Cocks—Trevor Parker

1965 (London)
The following ten positions were won in the first 40 birds of the Standard Class:

2nd E Griffiths
4th W Holmes
14th T J Davies
15th E Griffiths
16th T J Parker

18th E Griffiths
21st W Holmes
27th Baugh & Son
28th T J Parker
37th Mrs A & R Bowes

Team Results Standard Class (14 countries):
1. Great Britain
2. Czechoslovakia
3. Belgium

4. Holland
5. France

Best in Show (Best Hen): L Bohumil, Czechoslovakia.
Best Opposite Sex: Edgar Griffiths, Llanelly.

Sporting Class: 12 non-competitive national teams.

1967 (Vienna)
The following positions were won in the first ten cocks and hens in the Standard Class:

Hens:
1st T J Davies
2nd Holmes Bros
3rd T J Parker
4th T J Parker and T J Davies

Cocks:
2nd H L Webley
3rd T J Davies
4th T J Davies

Team Results Standard Class (16 countries):

1. Great Britain
2. Czechoslovakia
3. Holland

4. France
5. Belgium

Best in Show (Best Cock): L Bertreaux (France).
Best Opposite Sex: T J Davies.

1969 (Katowice)
The following positions were won in the first ten cocks and hens in the Standard Class:
Hens:
1st D Cardwell
2nd T J Davies

Cocks:
1st D Cardwell
5th Holmes Bros

Team Result (18 countries):
1. Holland
2. Great Britain
3. Czechoslovakia

Best in Show (Best Cock): D Cardwell.

1971 (Brussels)
The following positions were won in the first 40 in the Standard Class;
Hens:
1st Mr & Mrs Pugh
20th Holmes Bros

Cocks:
19th Mr & Mrs Pugh
38th H L Webley

1973 (Dusseldorf)
No individual awards were made other than Best Cock and Best Hen in International competition. Each country was awarded medals for Best Cock and Hen.
   Premier award in show, Cocks: T J Davies & Son, Llanelli; 2nd Best British Cock — Silver Medal (which for the other countries would be for the best Cock) T J Davies & Son.
   Silver Medal, Best British Hen, H L Webley, Llanelly Hill.

1975 (Budapest)
Team Result (15 countries):
1. Czechoslovakia
2. Holland
3. Great Britain

1977 (Blackpool)
Team Result (17 countries):
1. West Germany
2. Great Britain
3. Holland

Best Hen in Show: C Miller, Blackburn.

1979 (Amsterdam)
Team Result (18 countries):
1. Czechoslovakia
2. W Germany
3. Romania

Great Britain 7th

1981 (Tokyo)
Team Result (18 countries):
1. Holland
2. Romania
3. Czechoslovakia

Great Britain 11th

1983 (Prague)
Team Result (19 countries):
1. Czechoslovakia
2. Poland
3. Hungary

Great Britain 17th

1985 (Oporto)
Team Result (17 countries):
1. Czechoslovakia
2. W Germany
3. Hungary

Great Britain 15th

1987 (Dortmund)
Team Result (19 countries):
1. Czechoslovakia
2. W Germany
3. Romania

Great Britain 13th

1989 (Katowice)
Team Result (19 countries):
1. Czechoslovakia
2. Poland
3. Romania

Great Britain 16th

1991 (Verona)
Team Result (17 countries):
1. Holland
2. Portugal
3. Czechoslovakia

Great Britain 13th

INTERNATIONAL STANDARD

(Prepared during the FCI congress at Oporto, Portugal 1985 as modified at Katowice, Poland 1989)

MEANING AND OBJECT OF THE STANDARD — The Standard is intended to define the ideal pattern of bodily constitution of a fast and hardy homing pigeon. The description is the result of the knowledge acquired so far. It is, for the fancier, a means of achieving greater perfection and is at the same time used as a criterion for judging pigeons in exhibition.

1. THE HEAD — Should be convex, either round or slightly flat on the top. The back of the head should be deep. The forehead should be developed in width as well as in height. The head, which is finer and smaller for the female, should be more developed for the male.

2. THE BEAK — Must be in proportion to the head of the pigeon and be well closed.

3. THE NOSE OR WATTLE — Preferably, but having regard to the age of the pigeon, the wattle should not be over-developed. It must be dry and white.

4. THE EYES — Preferably the eyes should be well set, shine brightly and reflect the health of the pigeon; their colour is not important. The membranes of the eye, although not very important, must be well defined. Shades of grey or white are recommended.

5. NECK AND COLLAR — The neck should be thickly feathered and the collar as powerful as possible.

6. BALANCE AND MUSCLE STRUCTURE — The whole of the trunk should be very firm, well-balanced, of an aerodynamic shape, with a flexible muscle structure in proportion to the build of the pigeon. The small of the back must be well-proportioned and well-feathered, the back must be wide between the shoulders and should taper towards the tail so as to form a solid whole with the base of this tail. A well-proportioned chest width and a rounded line of the sternum are prerequisites to the balance of the trunk. The centre of gravity of the trunk is situated in the anterior part.

7. THE WINGS — The arm of the wing should be thick, short and thickly feathered. The joint to the body should be as short as possible.
When spread out, the wings should be slightly curved.
The secondary wing should be formed of feathers as wide as possible; as to the length of this secondary wing, it must be in proportion to the primary wing. The last four should be spaced out so as to facilitate flight. Their length should be proportional to the body, they must not be too wide at the base and must taper towards the end and be slightly rounded at the extremity.
The wings, and more particularly the quills of the primaries, should be as flexible as possible.
The small feathers of the wings should be abundant and the ones from the top should nicely overlap one another like the slates on a roof, and spread as far as possible towards the extremities of the primary and the secondary wings so as to reinforce the primaries. The small feathers from underneath the wing should be abundant and silky so as to enable a smooth flow of air during flight.

8. THE FORK — The bones of the fork should be firm and tend towards one another. Slight, thin, soft bones constitute a very big fault. A pelvis opening of 2 to 3mm when at rest is considered a very slight imperfection, if the bones of the furculum are strong and firm. These should be as close as possible to the sternum. These criteria will not be judged as severely for females as they will for males.

9. THE RUMP — This should harmoniously continue the line formed by a strong and proportioned small of the back area. It should be abundantly covered on all sides with fine, silky feathers, which should cover the feathers of the tail as much as possible.

10. THE TAIL, THE LEGS AND FEET — These must be in proportion with the body.

11. THE PLUMAGE — Should be rich, abundant and silky; the colour does not matter.

12. PRESENTATION — The true racing pigeon does not always 'show' itself very well; it is a nervous and restless creature and is more tense than an actual exhibition pigeon, which has straighter legs and holds its wings closer to the body. The eye of a racing pigeon is more piercing, it does not relax and it is attentive to everything that happens around it. The lack of pose in the appearance does not constitute a fault in this case.

13. THE WEIGHT — Experience has proved that there is only a small number of heavy pigeons which are awarded winners from all distances. On the contrary, there is a large number of pigeons which are small and light, and the majority of good subjects are of medium weight. The weight of a pigeon shall be in proportion to the body.

INTERNATIONAL STANDARD OF SPORTING PIGEON

JUDGEMENT — The point scale is closely linked to the verbal description of the quality.

	Max
Head, eye and expression, aesthetic aspect, judgement by eye	10
Balance and muscle structure	30
General bone structure, breast bone and rear fork	15
Back, rump and point of attachment	15
Wing, tail and quality of plumage	30
	100

REQUIREMENTS FOR ENTRY IN OLYMPIADS

The two teams, Standard and Sport, consist of ten pigeons each. It is no longer compulsory to participate in both categories.
a) STANDARD — The team of each country must consist of five males and five females, which must meet at least the conditions listed below:
TOTAL MILES / PRIZES IN THE TWO PRECEDING YEARS —
1,875 miles — males
1,562.5 miles — females
No duplications allowed. 30% of the qualifying distances must have been flown in the year preceding the Olympiad.
MILES/COMPETITION (MINIMUM DISTANCE PER RACE) 93.7 miles.
MINIMUM NUMBER OF PIGEONS PARTICIPATING PER RACE — 200 — 20% of prizes (or one prize per five).
b) SPORT (either males or females)
1. Each Federation taking part in the Olympiad will be allowed to select 12 pigeons for the sport category.
2. The pigeons will be divided into three categories:
   CATEGORY 1: pigeons with prizes between 100 and 400 km.
   CATEGORY 2: pigeons with prizes between 300 and 600 km.
   CATEGORY 3: pigeons with prizes above 500 km.
3. For the first category the pigeons must have won ten prizes for a minimum of 2,000 km.

For the second category the pigeons must have won eight prizes for a minimum of 3,000 km.

For the third category the pigeons must have won six prizes for a minimum of 4,000 km.

4. The prizes won will be per-five prizes, ie 20% of the participants.

5. The Federations can register their pigeons in the category they wish but a pigeon can be registered in one category only.

6. In order to be accounted for the competition will have to fulfil the following requirements:

a) rally at least 250 pigeons

b) include a minimum of 25 participants

7. The placing will be made by applying the following formula to each prize won:

$$\frac{\text{prize won} \times 1,000}{\text{number of participating pigeons}}$$

8. The pigeon with the smallest coefficient will be proclaimed winner of each category.

RECORD OF ACHIEVEMENT TABLES, WHICH WILL BE PLACED ON THE CAGES OF THE PIGEONS EXHIBITED — The participating Federations will receive from the organising Federation tables to be filled in for the pigeons selected for the Olympiad. These tables, duly completed and bearing the stamp of the Federation concerned, to vouch for the authenticity of the information given thereon, must be handed in to the representatives of the organising Federation when the pigeons are presented to be transferred into exhibition cages.

OFFICIAL RESULTS — Official results which justify the prizes for the pigeons of the STANDARD category shall have to be HANDED IN AT THE SAME TIME as the Record of Achievement tables, otherwise the pigeons shall not be included in the league table. To make identification of the placing of the pigeons presented easier, it is compulsory to underline them on the results, preferably in red ink. Considering that it is the organising Federation's responsibility to ensure the efficient organisation of the Olympiad, under Article 15 of the Statutes of the FCI, it is absolutely necessary that they know sufficiently in advance which Federation will participate in the Congress on the one hand, and in the Sport and Standard exhibitions on the other. This is why it is absolutely necessary that the decision of each Federation is communicated, without fail, to the Secretariat of the FCI, 39 Rue de Livourne, 1050 Brussels (Belgium) BEFORE the date specified by returning the two forms sent to each Federation, duly completed.

The three best-placed Federations in the International League Table — Standard category — will receive a trophy offered by the FCI. The owners of the pigeons, Standard and Sport categories, will be awarded a diploma as well as a souvenir medal.

COLOUR ABBREVIATIONS

Below is a selection of abbreviations which the fancier may find helpful when filling in his race sheet or nomination list. These are particularly useful when used in advertisements or in captions for illustrations.

Bk — Black
B — Blue
R — Red
Ch — Chequer
M — Mealy
Griz — Grizzle
Lt — Light
Dk — Dark
w/f — white flighted
Wh — White
P — Pied

Str — Strawberry (ie Str M)
Smokey — spelt out
Barless — spelt out
Pen — Pencil
Sil — Silver
Gay — spelt out
Y — Yellow
Mos — true Mosaic
Plum — spelt out
Velvet — spelt out
Opal — Opal Mosaic

Cock and Hen become C and H.

For a detailed article on abbreviations see *Racing Pigeon Pictorial*, Number 21.

PAIRS FOR COLOUR BREEDING

Pairing Cock × Hen	Youngsters Cocks	Hens
Blue × Blue	Blue	Blue
Blue × Mealy	Mealy	Blue
Blue × Blue Chequer	Blue, Blue Chequer	Blue, Blue Chequer
Blue × Red	Mealy, Red	Blue, Blue Chequer
Mealy × Blue	Blue, Mealy	Blue, Mealy
Mealy × Mealy	Mealy	Blue, Mealy
Mealy × Blue	Mealy, Red, Blue	Mealy, Red, Blue
Chequer	Blue Chequer	Blue Chequer
Mealy × Red	Mealy, Red	Mealy, Red, Blue, Blue Chequer
Blue Chequer × Blue	Blue, Blue Chequer	Blue, Blue Chequer
Blue Chequer × Mealy	Mealy, Red	Blue, Blue Chequer
Blue Chequer × Blue Chequer	Blue, Blue Chequer	Blue, Blue Chequer
Blue Chequer × Red	Mealy, Red	Blue, Blue Chequer
Red × Blue	Blue, Blue Chequer, Red, Mealy	Blue, Blue Chequer, Red, Mealy
Red × Mealy	Mealy, Red	Mealy, Red, Blue, Blue Chequer
Red × Blue Chequer	Blue, Blue Chequer, Mealy, Red	Blue, Blue Chequer, Mealy, Red
Red × Red	Mealy, Red	Mealy, Red, Blue, Blue Chequer

THE MATING, LAYING AND POSITION FOR RACING TABLE

The table on the page opposite cannot be guaranteed for all birds as there will be considerable variation depending on conditions, but is intended to give a rough guide for the less experienced fancier.

HOW TO FIND FROM DATE OF MATING

DATE OF LAYING ....Add 11
DATE OF HATCHING ....Add 29
DATE OF LAYING AGAIN ....Add 43

These figures should be added to the table number of the date of mating.

HOW TO FIND DATE OF MATING SO THAT THE BIRD IS IN THE RIGHT POSITION FOR RACING

Best Position	First Round	Second Round
Driving	5	38
Point of Lay	10	43
Sitting 3 days................	13	46
Sitting 7 days................	17	50
Sitting 10 days...............	20	53
Sitting 14 days..............	24	57
Hatching......................	28	61
Young in nest 3 days.......	31	64
Young in nest 7 days.......	35	68
Young in nest 10 days	38	71
Young in nest 14 days	42	75
Young in nest 21 days	49	82

These numbers should be *subtracted* from the table number of the race date — result gives table number of date of mating.

Example

Birds mated 6 April......................Table number 37
Date of laying again ....Add 43
     TOTAL ....80—19 May

Example

Date of race, 8 July ....Table number 130
Bird's best position in three-day old youngsters
                                                     Subtract 64
     Result: 130 minus 64 equals 66 — 5 May
Birds should be mated for previous round 5 May

March	Table No	April	Table No	May	Table No	June	Table No	July	Table No
1	1	1	32	1	62	1	93	1	123
2	2	2	33	2	63	2	94	2	124
3	3	3	34	3	64	3	95	3	125
4	4	4	35	4	65	4	96	4	126
5	5	5	36	5	66	5	97	5	127
6	6	6	37	6	67	6	98	6	128
7	7	7	38	7	68	7	99	7	129
8	8	8	39	8	69	8	100	8	130
9	9	9	40	9	70	9	101	9	131
10	10	10	41	10	71	10	102	10	132
11	11	11	42	11	72	11	103	11	133
12	12	12	43	12	73	12	104	12	134
13	13	13	44	13	74	13	105	13	135
14	14	14	45	14	75	14	106	14	136
15	15	15	46	15	76	15	107	15	137
16	16	16	47	16	77	16	108	16	138
17	17	17	48	17	78	17	109	17	139
18	18	18	49	18	79	18	110	18	140
19	19	19	50	19	80	19	111	19	141
20	20	20	51	20	81	20	112	20	142
21	21	21	52	21	92	21	113	21	143
22	22	22	53	22	83	22	114	22	144
23	23	23	54	23	84	23	115	23	145
24	24	24	55	24	85	24	116	24	146
25	25	25	56	25	86	25	117	25	147
26	26	26	57	26	87	26	118	26	148
27	27	27	58	27	88	27	119	27	149
28	28	28	59	28	89	28	120	28	150
29	29	29	60	29	90	29	121	29	151
30	30	30	61	30	91	30	122	30	152
31	31			31	92			31	153

TIME OF FLIGHT AND VELOCITY APPROXIMATOR

This table can be used to estimate the time of arrival from a race point if the average velocity is known. Example: Race Distance is 450 miles. Birds will make about 1,100 ypm. Bird can be expected 12 hours after liberation.

The table will also enable a fancier to have some idea of his velocity after timing in. Example: Race Distance is 250 miles. Bird timed in 6 hours 5 minutes after liberation, bird will make a velocity of about 1,200 ypm.

All times to nearest minute

Estimated Velocity Y.P.M.	50 miles		75 miles		100 miles		125 miles		150 miles	
	H	M	H	M	H	M	H	M	H	M
700	2	06	3	09	4	12	5	14	6	17
800	2	50	2	45	3	40	4	35	5	30
900	1	38	2	27	3	16	4	04	4	53
1,000	1	28	2	12	2	56	3	40	4	24
1,100	1	20	2	00	2	40	3	20	4	00
1,200	1	13	1	50	2	27	3	03	3	40
1,300	1	08	1	42	2	15	2	49	3	23
1,400	1	03	1	34	2	06	2	37	3	09
1,500	0	59	1	28	1	57	2	27	2	56
1,600	0	55	1	23	1	50	2	18	2	45

	175 miles		200 miles		225 miles		250 miles		275 miles	
700			8	23	9	26	10	29	11	32
800	6	25	7	20	8	15	9	10	10	05
900	5	42	6	31	7	20	8	09	8	58
1,000	5	08	5	52	6	36	7	20	8	04
1,100	4	40	5	20	6	00	6	40	7	20
1,200	4	17	4	53	5	30	6	07	6	43
1,300	3	57	4	31	5	05	5	38	6	12
1,400	3	40	4	11	4	43	5	14	5	46
1,500	3	25	3	55	4	24	4	53	5	23
1,600	3	13	3	40	4	08	4	35	5	03

	300 miles		325 miles		350 miles		375 miles		400 miles	
	H	M	H	M	H	M	H	M	H	M
700	12	34	13	37	14	40	15	43	16	46
800	11	00	11	55	12	50	13	45	14	40
900	9	47	10	35	11	24	12	13	13	02
1,000	8	48	9	32	10	16	11	00	11	44
1,100	8	00	8	40	9	20	10	00	10	40
1,200	7	20	7	57	8	33	9	10	9	47
1,300	6	46	7	20	7	54	8	28	9	02
1,400	6	17	6	49	7	20	7	51	8	23
1,500	5	52	6	21	6	51	7	20	7	49
1,600	5	30	5	58	6	25	6	53	7	20

	425 miles		450 miles		475 miles		500 miles		525 miles	
700	17	49	18	52	19	54	20	57	22	00
800	15	35	16	30	17	25	18	20	19	15
900	13	51	14	40	15	29	16	18	17	06
1,000	12	28	13	12	13	56	14	40	15	24
1,100	11	20	12	00	12	40	13	20	14	00
1,200	10	23	11	00	11	37	12	13	12	50
1,300	9	35	10	09	10	43	11	17	11	51
1,400	8	54	9	26	9	57	10	29	11	00
1,500	8	19	8	48	9	17	9	47	10	16
1,600	7	48	8	15	8	43	9	10	9	38

METRIC SPEED & DISTANCE CONVERSION

TO CONVERT YPM TO MPM:
Read down central column (B) for number of YPM number of MPM in column (C).

TO CONVERT MPM TO YPM:
Read metres in central column (B) and equivalent yards in column A.

TO CONVERT MILES TO KILOMETRES:
Read down central column (B) for number of miles; number of kilometres in first column (A).

TO CONVERT KILOMETRES TO MILES:
Read kilometres in central column (B) and equivalent miles in column C.

MPM to YPM		YPM to MPM
A	B	C
651	600	549
706	650	594
760	700	640
814	750	682
868	800	728
923	850	774
977	900	823
1031	950	867
1086	1000	914
1140	1050	960
1194	1100	1006
1248	1150	1052
1303	1200	1097
1357	1250	1143
1411	1300	1189
1466	1350	1234
1520	1400	1280
1547	1450	1326
1630	1500	1372
1737	1600	1463
1846	1700	1554
1954	1800	1646

All figures correct to nearest whole number
(For exact calculations multiply YPM x ·9144 for MPM; to convert MPM x 1·0856 for YPM)

Miles to Kilometres		Kilometres to Miles
A	B	C
8	5	3
16	10	6
32	20	12
48	30	19
64	40	25
80	50	31
97	60	37
113	70	43
129	80	50
145	90	56
161	100	62
240	150	93
322	200	124
402	250	155
483	300	186
644	400	249
805	500	311
966	600	373
1127	700	435
1287	800	497
1448	900	560
1609	1000	621

All figures correct to nearest whole number
(For exact calculations multiply miles x 1·6093; for kilometres; multiply kilometres x ·62137 for miles)

METRIC TIME OF FLIGHT AND VELOCITY APPROXIMATOR

This table can be used in the same way as the non-metric table but shows distance in kilometres to metres per minute.

Distance (km)	Metres per minute										
	1800	1700	1600	1500	1400	1300	1200	1100	1000	900	800
	H.m.s.	H.m.s.	H.m.s.	H.m.s.	H.m.s.	H.m.s.	H.m.s.	H.m.s.	H.m.s.	H.m.s.	H.m.s.
70	0.38.54	0.41.10	0.43.45	0.46.40	0.50.00	0.53.50	0.58.20	1.03.38	1.10.00	1.17.46	1.27.30
80	0.44.27	0.47.03	0.50.00	0.53.20	0.57.08	1.01.32	1.06.40	1.12.43	1.20.00	1.28.53	1.40.00
90	0.50.00	0.52.55	0.56.15	1.00.00	1.04.17	1.09.14	1.15.00	1.21.49	1.30.00	1.40.00	1.52.30
100	0.55.33	0.58.48	1.02.30	1.06.40	1.11.25	1.16.56	1.23.20	1.30.54	1.40.00	1.51.06	2.05.00
110	1.01.06	1.04.41	1.08.45	1.13.20	1.18.34	1.24.38	1.31.40	1.40.00	1.50.00	2.02.13	2.17.30
120	1.06.40	1.10.34	1.15.00	1.20.00	1.25.42	1.32.19	1.43.00	1.49.05	2.00.00	2.13.20	2.30.00
130	1.12.13	1.16.27	1.21.15	1.26.40	1.32.51	1.40.00	1.48.20	1.58.11	2.10.00	2.24.26	2.42.30
140	1.17.46	1.22.20	1.27.30	1.33.20	1.40.00	1.47.41	1.56.40	2.07.16	2.20.00	2.35.33	2.55.00
150	1.23.20	1.28.13	1.33.45	1.40.00	1.47.09	1.55.23	2.05.00	2.16.22	2.30.00	2.46.40	3.07.30
160	1.28.53	1.34.06	1.40.00	1.46.40	1.54.17	2.03.04	2.13.20	2.25.27	2.40.00	2.57.46	3.20.00
170	1.34.26	1.40.00	1.46.15	1.53.20	2.01.25	2.10.45	2.21.40	2.34.33	2.50.00	3.08.53	3.32.30
180	1.40.00	1.45.51	1.52.30	2.00.00	2.08.33	2.18.26	2.30.00	2.43.38	3.00.00	3.20.00	3.45.00
190	1.45.33	1.51.44	1.58.45	2.06.40	2.15.41	2.26.08	2.38.20	2.52.43	3.10.00	3.31.06	3.57.30
200	1.51.06	1.57.37	2.05.00	2.13.20	2.22.49	2.33.50	2.46.40	3.01.49	3.20.00	3.42.13	4.10.00
225	2.05.00	2.12.18	2.20.37	2.30.00	2.40.42	2.53.04	3.07.30	3.24.32	3.45.00	4.10.00	4.41.15
250	2.18.53	2.27.00	2.36.15	2.46.40	2.58.34	3.12.18	3.28.20	3.47.16	4.10.00	4.37.46	5.12.30
275	2.32.45	2.41.43	2.51.52	3.03.20	3.16.25	3.31.32	3.49.10	4.10.00	4.35.00	5.05.30	5.43.45
300	2.46.39	2.56.26	3.07.30	3.20.00	3.34.17	3.50.46	4.10.00	4.32.43	5.00.00	5.33.17	6.15.00
325	3.00.33	3.11.10	3.23.07	3.36.40	3.52.08	4.10.00	4.30.50	4.55.27	5.25.00	6.01.04	6.46.15
350	3.14.27	3.25.52	3.38.45	3.53.20	4.10.00	4.29.13	4.51.40	5.18.11	5.50.00	6.28.50	7.17.30
375	3.28.20	3.40.34	3.54.22	4.10.00	4.27.51	4.48.27	5.12.30	5.40.54	6.15.00	6.56.37	7.48.45
400	3.42.13	3.55.17	4.10.00	4.26.40	4.45.42	5.07.41	5.33.20	6.03.38	6.40.00	7.24.24	8.20.00
425	3.56.07	4.10.00	4.25.37	4.43.20	5.03.34	5.26.55	5.54.10	6.26.21	7.05.00	7.52.12	8.51.15
450	4.10.00	4.24.41	4.41.15	5.00.00	5.21.25	5.46.09	6.15.00	6.49.03	7.30.00	8.20.00	9.22.30
475	4.23.53	4.39.20	4.46.52	5.16.40	5.39.17	6.05.23	6.35.50	7.11.48	7.55.00	8.47.46	9.53.45
500	4.37.47	4.54.00	5.12.30	5.33.20	5.57.08	6.24.36	6.56.40	7.34.32	8.20.00	9.15.33	10.25.00
525	4.51.40	5.08.42	5.28.08	5.50.00	6.15.00	6.43.50	7.17.30	7.57.16	8.45.00	9.43.20	10.56.15
550	5.05.33	5.23.25	5.43.45	6.06.40	6.32.51	7.03.04	7.38.20	8.20.00	9.10.00	10.11.06	11.27.30
575	5.19.25	5.38.07	5.59.22	6.23.20	6.50.42	7.22.18	7.59.10	8.42.44	9.35.00	10.38.53	11.58.45

TABLE OF COMPARATIVE VALUES OF FOODS

Feeding stuff	Dry Matter	Protein Crude	Digestible	Carbohydrates Soluble	Digestible	Oil Ether Extr	Digestible	Fibre
	%	%	%	%	%	%	%	%
Legumes								
Maple Peas	86.0	22.5	19.4	53.7	49.9	1.6	1.0	5.4
Tic beans	85.7	27.0	23.0	50.2	46.2	1.5	1.2	7.8
Dun beans								
Tares	86.7	26.0	22.9	49.8	45.8	1.7	1.5	6.0
Vetches								
Soya beans	90.0	33.2	29.5	30.5	20.8	17.5	15.8	4.1
Cereals								
Maize (and maize meal)	87.0	9.9	7.1	69.2	65.7	4.4	3.9	2.2
Wheat (strong varieties)	87.9	13.0	11.4	69.6	64.0	2.2	1.2	1.6
Wheat (weak varieties)	88.5	9.5	8.4	73.8	67.8	2.0	1.2	1.7
Barley (and barley meals)	85.1	8.6	6.5	67.9	62.2	1.5	1.2	4.5
Oats	86.7	10.3	8.0	58.2	44.8	4.8	4.0	10.3
Dari (Kafir corn) (Sorghum grain)	88.9	9.6	7.7	71.2	60.5	3.8	3.0	1.9
Buckwheat	85.9	11.3	8.5	54.8	42.3	2.6	1.9	14.4
Rice (polished)	87.4	6.7	5.8	78.0	75.8	0.4	0.2	1.5
Oil Seeds								
Linseed	92.8	24.2	19.4	22.9	18.3	36.5	34.7	5.5
Hemp	91.1	18.2	13.7	21.1	16.8	32.6	29.3	15.0
Sunflower seed	92.5	14.2	12.8	14.5	10.3	32.3	30.7	28.1
Ingredients for Pellets								
Wheat middlings (fine)	87.3	15.7	13.2	64.0	52.0	3.4	3.0	1.8
Wheat middlings (coarse)	86.5	16.4	13.8	56.2	45.5	5.0	4.3	5.3
Wheat pollards	86.7	14.3	11.6	55.6	44.5	4.8	4.0	7.7
Wheat bran	86.4	13.5	10.6	53.0	38.0	3.9	2.8	10.6
Maize (germ meal)	89.3	13.0	10.4	55.1	45.8	13.5	12.8	4.1
Maize gluten feed	89.6	23.5	20.0	56.7	49.3	3.4	2.7	3.5
Maize gluten meal	90.9	35.5	30.6	47.5	42.6	4.7	4.4	2.1
Oat feed (high grade)	91.0	10.3	—	53.7	—	5.0	—	17.5
Oat feed (ordinary)	92.0	5.5	—	52.5	—	2.4	—	27.6
Oat husks	94.0	2.0	—	54.0	19.4	1.0	0.4	33.0
Brewers' Grain dried	89.7	18.3	13.0	45.9	27.6	6.4	5.6	15.2
Soya bean meal (extracted)	88.7	44.7	40.3	31.9	24.7	1.5	1.4	5.1
Palm kernel meal (extracted)	90.0	19.0	17.1	49.0	43.5	2.0	1.9	16.0
Ground nut cake meal (decorticated)	89.7	46.8	37.6	23.2	18.6	7.5	6.7	6.4
Sunflower cake	90.4	37.4	33.6	20.4	14.6	13.8	12.2	12.1
Linseed cake meal (English)	88.8	29.5	25.3	35.5	28.5	9.5	8.7	9.1
Linseed cake (foreign)	89.0	32.3	27.8	32.2	25.8	9.9	9.1	8.7
Linseed meal (extracted)	88.2	35.7	30.8	33.9	27.2	3.1	2.8	9.0
Biscuit crumbs	90.6	13.9	12.5	74.6	68.5	0.7	0.6	0.3
Vegetables and Wild Seeds								
Acorns (fresh)	50.0	3.3	2.7	36.3	36.6	2.4	1.9	6.8
Acorns (dried)	85.0	5.7	4.6	61.6	55.5	4.1	3.3	11.6
Beech mast				(Approx as Linseed)				
Lucerne (in flower)	24.0	3.9	2.7	9.2	5.7	0.8	0.4	7.8
Alfalfa (lucerne) meal	89.3	14.6	9.2	33.5	22.0	1.8	0.9	31.0
Cabbage	11.0	1.5	1.1	5.9	4.6	0.4	0.2	2.0
Kale	14.8	2.5	1.8	8.7	7.0	0.3	0.2	1.7
Potatoes	23.8	2.1	1.1	19.7	17.7	0.1	—	0.9
Potato peelings (hand)	21.2	2.1	1.6	17.0	16.4	0.1	—	0.7
Yeast (dried)	95.7	48.5	41.6	35.5	29.2	0.5	0.2	0.5
Meat and Fish Products								
Fish meal	87.0	55.6	50.0	2.1	—	4.4	4.2	—
Meat meal	89.2	72.2	55.7	—	—	13.2	12.5	—
Meat and bone meal	90.8	50.5	39.0	—	—	10.0	9.4	—
Blood meal	86.0	81.0	72.7	1.5	—	0.8	0.8	—
Milk, fresh (whole)	12.8	3.4	3.2	4.8	4.8	3.9	3.9	—
Milk, whole (dried)	95.8	25.5	24.0	37.4	37.4	26.5	26.5	—
Milk (separated)	9.4	3.5	3.3	5.0	5.0	0.1	0.1	—
Milk (skimmed dried)	89.8	32.8	29.6	48.8	48.8	0.3	0.3	—
Buttermilk (dried)	90.0	42.3	34.5	24.3	24.3	11.2	11.2	—
Whey (dried)	92.2	12.6	—	70.5	—	1.4	—	—

OSMAN MEMORIAL CUPS

These eight cups were purchased from donations as a memorial to Lt-Col A H Osman, OBE, a founder of the sport and first editor of The Racing Pigeon. They are awarded by an independent panel of judges for the eight best performances of the year; four for pigeons flying North and four for pigeons flying South.

1931
South Road
BARR BROS, Ballymena
DRYDEN COCHRANE, South Shields
W STUBBS, Cosham
MAJOR W JAMES, Wolverhampton
North Road
CAPT A H QUIBELL, Lincoln
P HARWOOD, Newmarket
A BEGBIE, Bristol
H A BRIDGE, Rayleigh
1932
South Road
R M ANTINGHAM, Oxford
W BASTON, Alnwick
J HARDAKER, Leeds
RECORD & TALLMAN, Exeter
North Road
A BEGBIE, Bristol
A MILES, Tottenham
F STEVENSON, Skegness
G WOODCOCK, Brentwood
1933
South Road
ELLIOTT BROS, London
H K JACKSON, Brackley, Northamptonshire
JOHNSTON & BROWN, Sunderland
WADSWORTH & YATES, Walkden
North Road
ARGENT BROS, London
BRANSTON, SON & SMITH, Mansfield
CHIVERS BROS, Tunley, Devon
MARCH BROS, Bedminster
1934
South Road
COBBLEDICK & WALLACE, Ashington
FRAME & COCHRANE, Strathaven, Lanarkshire
V ROBINSON, Woolston, Hampshire
J T CLARK, Windermere
North Road
S R ATKINS, Ipswich
S DAY, Spalding
W SAVAGE, Stanmore, London
J HICKS, Bristol
1935
South Road
SIR WILLIAM JURY, Reading
CASEY BROS, Strathaven, Lanarkshire
F W PELAN, Lisburn, Northern Ireland
W WHARTON, Liverpool
North Road
J R MARRIOTT, Derby
CORT & SON, Leicester
R W WALKER, Bristol
BUTLER BROS, Clapton Park, London
1936
South Road
R TUSTIAN, Oxford
A HALLIMOND, Deaham, Durham
J RUSSELL, Quarter, Lanarkshire
R ALLEN, Amble, Morpeth
North Road
CAPT J S THOMPSON, Denham, Buckinghamshire
J WALLINGTON, Rickmansworth, Hertfordshire
J H TAYLOR, Leicester
E H LULHAM, Chingford, Essex
1937
South Road
W H HARDCASTLE, Birmingham

J WARD, New Seaham, Durham
H GARDNER, Galgate, Lancashire
PRESCOTT BROS, Liverpool
North Road
R F TOWLE, Totton, Nottinghamshire
E W COX, Kenton, Middlesex
W BROWN, Bow, London
G H SOAR, Colchester
1938
South Road
SMITH & AITKEN, Lanark
H PARK, Armadale, West Lothian
ROWE & ROWE, Haswell
W HAYTON, Houghton-le-Spring
North Road
MAJOR G S ROGERS, Sheringham
W W COX, Kenton, Middlesex
WILLIAMS BROS, Rhymney
J MORRISON, Tottenham, London
1939
South Road
ANDERSON BROS, Falkirk, Stirlingshire
T CHALLINOR & SON, Chester
G FRANCIS, West Lothian
W HEYDON, Seaham, Durham
North Road
BEGBIE & POWELL, Bristol
JEPSON BROS & CURTIS, Peterborough
MORGAN BROS, Llanhilleth, Monmouthshire
L C PALMER, Croxley Green, Hertfordshire
1940
South Road
BROWN BROS, Symington, Lanarkshire
J FEENEY, Winchburgh, West Lothian
J CAMERON, Blantyre, Lanarkshire
H BROCOTT, Stoke-on-Trent
North Road
BRIGHT & DOWDALL, Newport, Gwent
P STANDERWICK, Bridgwater, Somerset
C BROWN, Bournemouth
MRS R JOHNSON, Mickleover, Derby
1941
South Road
L WAINWRIGHT, Cheadle, Staffordshire
T A WARRINGTON, Prestbury, Cheshire
W SNEDDON, New Lanark
J T WHITTAKER, Clayton-le-Moors
North Road
T JENKINS, Abertillery
L C PALMER, Croxley Green, Hertfordshire
R S FROST, Luton, Bedfordshire
T TILL, Rugeley, Staffordshire
1942
South Road
TIMMIS BROS, Audley, Staffordshire
J WRIGHT, Cheadle, Staffordshire
G WHEAT, Cheadle, Staffordshire
S BARNES, Oakham, Dudley
North Road
E ROWBOTHAM, Stroud, Gloucester
F HODDER, Newport, Gwent
G A LOVELL, Hatfield, Hertfordshire
SMITH & SANKEY, Droitwich
1943
South Road
T CARTWRIGHT, Rugeley, Staffordshire
E STEVENSON, Mansfield
A L HIGBY, Chesham, Buckinghamshire
SHAKESPEARE & SON, Oldbury, Worcester

1944
South Road
C HOMER, Stowbridge
T H SILLITO, Stoke-on-Trent
W CAMPBELL, Cheadle, Staffordshire
T MORGAN, Airdrie, Lanarkshire
North Road
LORD CAWLEY, Leominster
HOWELL & PITCHER, Pontywain, Monmouthshire
WYATT BROS, Bristol
P STANDERWICK, Bridgwater, Somerset
1945
South Road
REITH BROS, Airdrie, Lanarkshire
W A BREAKS, Clitheroe, Lancashire
J MORGAN, Airdrie, Lanarkshire
A H RUSSELL, New Cross, London
North Road
H S PEARCE, Brentwood, Essex
JEPSON BROS & CURTIS, Peterborough
A J ASHBY, Croxley Green, Hertfordshire
G ALLNUTT, Chalfont St Peter, Buckinghamshire
1946
South Road
A RAWSON, Worksop
L GILBERT, Forest Hill, London
D T ANGUS, Symington, Lanarkshire
DAVIES BROS, Aberdare
North Road
C W ALLBUTT, Hendon, London
MRS R TWEDDELL, Tottenham, London
H NORFOLK, Guiseley, Yorkshire
J STANDERWICK, Bridgwater, Somerset

1947
South Road
T PILMOUR, Hartlepool
N CAMPBELL, Dunbar, Scotland
J HUNT, Atherton, Manchester
G E JARVIS, Eastry, near Sandwich
North Road
B WOOD, Acton Green, London
SLADE BROS, Enfield
S JARVIS, Bridgwater, Somerset
I GARDENER, Pontywain
1948
South Road
J SMITH & SON, Bolton, Lancashire
GOLDSTRA & SERGEANT, Milton, Staffordshire
TAYLOR & FORSHAW, Tarleton, Lancashire
M WARD, Blackhall, Durham
North Road
S E BIRCH, Crumlin, Monmouthshire
P STANDERWICK, Bridgwater, Somerset
J REES, Street, Somerset
A BUSH, Tibshelf, Derby
1949
South Road
F G TIPPER, Stoke-on-Trent
F H JARVIS, Harpenden, Hertfordshire
A BILVERSTONE, North Shields
C H BETTISON, Huthwaite, Nottinghamshire
North Road
P STANDERWICK, Bridgwater, Somerset
W G ANDERSON, Waltham Cross, Hertfordshire
W T CARR, Tottenham, London
FURLEY BROS, Blackwood, Monmouthshire
1950
South Road
MOORE & WOOTTON, Stoke-on-Trent
R SEAMAN, Cudworth, Barnsley
T D DIXON, Askam-in-Furness
W S LUND, Bolton, Lancashire
North Road
E STEVENSON, Mansfield
C WICK, Hatfield, Hertfordshire
J CHANNING & SON, Newport, Monmouthshire

D TINGEY, St Neots, Huntingdonshire
1951
South Road
J MACKAY, Annan, Dumfriesshire
F GODFREY, Worksop, Nottinghamshire
POTTS BROS, Stoke-on-Trent
SCOTT BROS, Isleworth, Middlesex
North Road
G J W TITMUSS, Wheathampstead, Hertfordshire
W MAYLOTT, Port Talbot, Glamorgan
ATWELL BROS, Newport, Monmouthshire
H J CHANDLER, Chalfont St Peter, Buckinghamshire
1952
South Road
MOORE & WOOTTON, Stoke-on-Trent
W W SMILES, Leatherhead
J MARSHALL, Hamilton, Lanark
TINKLER, SON & PIGGFORD, Co Durham
North Road
V DIVIT, Mansfield
C KELLAWAY & SON, Mill Hill
A E BAKER, Wood Green, London
POUNDSBERRY BROS, Bristol
1953
South Road
BARKER BROS, Hucknall, Nottinghamshire
W J EDWARDS & SON, Marlow, Buckinghamshire
H EVANS, Rowley Regis, Staffordshire
MONTGOMERY BROS, Catrine, Ayrshire
North Road
Mr & Mrs BENNETT, Ravenstone, near Leicester
L DAVIES, Ware, Hertfordshire
H HARDING, Spalding, Lincolnshire
C WARDLE & SON, Braunstone, Leicester
1954
South Road
D FOWLER, Methil, Fife
H JAMES, Jarrow
RANABOLDO BROS, W Molesey
TINKLER, SON & PIGGFORD, W Hartlepool
North Road
Mrs H A BRIDGE, Thundersley
E L HUNT, Leicester
T BETSON & SON, W Croydon
PHILLIPS & THOMAS, Tondu, S Wales
1955
South Road
J HODGSON, Drem
V ROBINSON, Southampton
HENRY BROS, Seaham
HOWDEN & JOHNSON, Castleford
North Road
P NUTT, Street, Somerset
J MASON & SON, Chatteris
ELLETT & CRONIN, Brixton
TATE BROS, Chiswick
1956
South Road
J W LANGSTONE, Worcester
J McGILLIVRAY & SON, Forth
R MUNDAY, Bolton
HEYDON BROS, Dawdon
North Road
T PHILLIPS, Penrhiwceiber
S JONES, Barry
H HATCHETT, Chesham
F HOLBARD, Tottenham
1957
South Road
E ROBINSON & PARTNERS, Fatfield
H SCOTT, Markinch
J KILGOUR, Salsburgh
G HAY, Bo'ness
North Road
E BURKE, Bow, London
J HUDSON, Ipswich
C GILLUM & JONES, Pontypool

Mr & Mrs CARTMELL, Mill Hill
1958
South Road
S V HOWELLS, Ferryhill
DUFF BROS, Easington Colliery
PINNEY BROS, Yeovil
J McGILLIVRAY & SON, Forth
North Road
S G BISS, Cheshunt
Mr & Mrs GARRAWAY, Harrow
JONES BROS, Bleanavon
A HUGHES, Bow, London
1959
South Road
A MARSHALL, Whitley Bay, Northumberland
H J KING, Blackwater, Camberley, Surrey
J & C SMILLIE, Fauldhouse, W Lothian
J A MacPHERSON, Brechin, Angus
North Road
S G BISS, Cheshunt, Hertfordshire
G TITMUSS, Wheathampstead, Hertfordshire
Mr & Mrs BENNETT, Ravenstone, Leicester
G K GARRETT, Harpenden, Hertfordshire
1960
South Road
F E NAYLOR, Thelwall, near Warrington
J HARDING, Tattenhall, near Chester
G R POUNDER, South Shields
D C TURNER, Lovedean
North Road
R HENDRICKSON, Roath, Cardiff
A E CARTER & SON, Peterborough
J CHANNING & SONS, Newport
A E BAKER, Wood Green
1961
South Road
J & P DORA, Arbroath
KILNER & DONKIN, Horden
J TAYLOR, Laughton
BRIDGE BROS, Burscough
North Road
AVIS & SON, Gospel Oak
JONES BROS, Cwngwrach
E F SOAMES, Manningtree
T MALES, Hitchin

1962
South Road
DOUGLAS & LITTLE, Chevington Drift
R TONG, Oxford
D ANGUS, Symington
C THOMPSON, Alnwick
North Road
T HUMBERSTONE, Kilburn
JONES BROS, Tredegar
G NUTT, Street
A STRANGE, Ipswich
1963
South Road
J LANCASHIRE, Aylesham
D SMITH, Great Ayton
A SHEPHERD, Arbroath
T JONES & SON, Rugeley
North Road
Mr & Mrs BAILEY, Harlow
ATWELL BROS, Newport
GWILLYM BROS & COOKSEY, Nantyglo
WENT BROS, Enfield, Middlesex
1964
South Road
J O WARREN & SON, Banks, near Southport
V PREDDY, High Littleton, near Bristol
J DREW, Freshwater, IoW
N B YOUNG & SONS, Broughty Ferry
North Road
J MORGAN, Abergavenny
T J PARKER, Treherbert

S & D CALKIN, High Barnet
G TITMUSS, Wheathampstead, Hertfordshire
1965
South Road
BEGLIN BROS, Bo'ness, W Lothian
W T GARDINER, Lockerbie
S J BANFIELD, Ensbury Park, Bournemouth
MOSSOP & MACKIE, Whitehaven
North Road
THROWER BROS, Stepney, London, E1
F D ALCOCK, Abercynon, Glamorgan
J H LOVELL, Donnington, near Spalding
WENT BROS, Enfield, Middlesex
1966
South Road
J CLARK, Consett, Co Durham
M NASH & SONS, Croy, near Kilsyth
WILLCOX BROS, Clutton, Bristol
J HAMILTON, Kennoway, Fife
North Road
E E STEVENS, Redbourn, Hertfordshire
WENT BROS, Enfield, Middlesex
BALE BROS & GRIFFITHS, Monmouthshire
JONES BROS, Cwmgwrach, Neath, Glamorgan
1967
South Road
J McLAREN, Cowplain, Portsmouth
L GRAHAM, Epsom, Surrey
R P EGGLESTON, Appleby, Westmorland
F CHEETHAM, Upton, Pontefract
North Road
G J W TITMUSS, Wheathampstead, Hertfordshire
T BASS & SON, Leyton, London
R EMRYS-JONES, Newport, Monmouthshire
A SOAR, Kingsbury, London
1968
South Road
BILLY ERWIN, Ballymena
J WILLIAMS & SON, Gilmerton
J HIBBERT, Warrington
ORLEY & SON, Hartlepool
North Road
G W J TITMUSS, Wheathampstead
R EMRYS-JONES, Sully
F CHANNING & SON, Newport
W BUTTON, Ipswich
1969
South Road
N CORRY, Dunmurry, Co Antrim
J MILLAR, Broxburn, Scotland
R BURTON, Harlepool, Co Durham
L BROWNSEY, near Stockport, Cheshire
North Road
J EDWARDS, Newport, Monmouthshire
WENT BROS, Enfield, Middlesex
K OTTON, Street, Somerset
J DING, Cambridge
1970
South Road
BERWICK & ELBECK, Whitehaven
A BRUCE, Fraserburgh
W CHALLEN, Cookham Dean, Berkshire
KNOX BROS, Dunbar
North Road
J & R BRILL, Ipswich
Mr & Mrs CROOKBAIN, Essex
G NAYLOR, Leytonstone
A G PLATT & SONS, Dagenham
1971
North Road
R ELLIOT, Dover
DAVIES, MOULDS & WILLIAMS, Pontygwaith
F SKOULDING & SON, March
FEAR BROS, Pontypridd
South Road
H THACKRAY, Leeds
J & J G PALEY, Keighley

C & H GREGORY, Radstock
R & M GRAHAM, Danderhall
1972
North Road
F DUNNING, Loughton
W SIMONS, Stevenage
J LOVELL, Donnington
J THOMAS & SON, Merthyr
South Road
L ADAMS, Bournemouth
T BOWLES, Austerfield
D & H PHILP, Hayle
C R MEDWAY, Southampton
1973
North Road
BOWEN BROS, Porthcawl
W A LAWSON, Ravenstone
R BLACK, Shelford
NORTHROP BROS, Ilford
South Road
J ALLFREE & JANE, Stanfre
W & D REE, Dundee
R CHURCHILL, Weymouth
J McGOWAN & KIRKBRIDE, Washington
1974
North Road
R DAVIES, Port Talbot
Mr & Mrs CLAYTON, Belvedere
R NEWSTEAD, Huntingdon
Mr & Mrs BARTON, Gillingham
South Road
E DISHINGTON, Reston
A SIMPSON & SON, Randalstown
A MARK, Thornton Heath
T HODGES, Taunton
1975
North Road
C BULLED, Harlow
K REYNOLDS, Swansea
R BURNS, Potters Bar
S & W THOROGOOD, Braintree
South Road
T W HOPEWELL, St Boswells
W MATHER, Marton
SHELL & SON, Alnwick
T DODD, Taunton
1976
North Road
C & K ROWE, Rickmansworth
C CHANNING, Newport, Gwent
R BRUNT, Newark
E FRENCH, Ipswich
South Road
A PAYNE, Keynsham
W MASSON, Inverallochy
J BORROWDALE & SON, Washington
W MATTHEWS, Castleford
1977
North Road
Mr & Mrs COWARD-TALBOTT, Chelmsford
S E JONES, Enfield
WILLIAMS BROS, Tredegar
H JONES, Gilfach Goch
South Road
R S WHYTE, Fraserburgh
RICHARDSON & SON, Seaton Delaval
W J BRADFORD, Sutton
P J ENTWHISTLE, Aylesbury

1978
North Road
G WILSON & SON, Wickford
BATES, SON & BURLEY, Langley Mill
P HULBERT, Newbridge-on-Wye
T CULVERHOUSE, Treorchy
South Road
R VAN ACHTER, Chester-le-Street

A H BENNETT, Church Stretton
E MOORHEAD & SON, Hartlepool
S G BISS, Norwich
1979
North Road
Mr & Mrs L MASSEY, Cheshunt
K & J THORPE, Colchester
Mr & Mrs E KIMBER, Sileby
E G EMMETT, Fordingbridge
South Road
R GILLETT, Southampton
J McKINNON, Ladybank
GALLOWAY & SON, Hartlepool
NEWCOMBE BROS, Macmerry
1980
North Road
T M CULVERHOUSE, Treorchy
JONES BROS & SON, Blaenavon
G WALKER, Harlow
K OTTON, Street
South Road
R SIMPSON, Mayfield
D SMITH, Great Ayton
B CLARYBURNS, Pontefract
G BYRNE, SONS & DAUGHTER, Kingston
1981
North Road
J TURNELL, Cheshunt
BOWEN BROS, Porthcawl
BEVAN & ROWLANDS, Pontypool
N C NELMES, Lydney
South Road
A MULHOLLAND & SON, Bristol
R SHERRATT & SON, Burton-on-Trent
BASLEY & ROBERTS, Horden
G A KITE, Banbury
1982
North Road
T M CULVERHOUSE, Treorchy
E J TOVEY, Pontypool
G R THOMAS, Merthyr Tydfil
P McKEOWN & SON, Cheshunt
South Road
A H BENNETT, Church Stretton
PUGH BROS, Wolverley
J C SANDERSON, Pathhead Ford
G HUNT & SON, Ash, Nr Canterbury
1985
North Road
J HAMMOND, Sutton Bridge
J CROSSAN, Downpatrick
W GILCHRIST, Cwmbran
T D WILSON, Bishop's Stortford
South Road
G W KIRKLAND, Coalport
Mr & Mrs D HARRINGTON, Lancaster
DONALDSON & LITTLE, Cochenzie
L A PEART & SON, Bradford-on-Avon
1986
North Road
M DEVEREUX, Grays
PREECE BROS, Cwmtillery
K DARLINGTON, Barry
C ADCOCK, Stapleford
South Road
WOODHOUSE BROS, Isle of Wight
GATES & SONS, Bolden
CALLAGHAN & SON, Fauldhouse
Mr & Mrs GODDARD & SON, Bradford
1987
North Road
K PURCELL, Cardiff
BATES, SON & BURLEY, Nottingham
R PRESTON, Ipswich
S J K JUDE & PROSSER, Cwmbran
South Road
CAMPBELL & WILSON, Witton Park

P NEE & DAUGHTER, Stokesley
G T BURGESS, Wraysbury
J LANE, Bromley
1988
North Road
J G DANIELS, Swansea
H BLAKER, Canning Town
Mr & Mrs C EVERARD, Skegness
Mr & Mrs W COOK, Abertillery
South Road
F & J GRAY, Newbiggin
W STEPHENSON, Shaftsbury
Mr & Mrs W GODDARD & SON, Bradford
G T BURGESS, Wraysbury
1989
North Road
ATWELL BROS, Cwmbran
J H LOVELL, Canwick
D MILDEN & SON, Plymstock
Mr & Mrs CAPPER, Gravesend
South Road
P SUMMERS & PTR, Selby
GUNN & CHERRIE, Roslin
D DELEA, Rainham

G SATTERLY & SON, Brockley
1990
North Road
E F CATCHPOLE, Bristol
JONES & WINSTONE, Pontypool
J SMALE, Abergavenny
H TAYLOR & SONS, Nottingham
South Road
D SANDERSON, Burnley
Mr & Mrs SEATON & SON, Guisborough
G WILLIS, Portadown
Mr & Mrs WRIGHT & DAUGHTER, York
1991
North Road
FISHLOCK & HARDING, Cardiff
J COTTERILL, Boughton, Notts
V SHAW & SON, Harlow
T HALEY, Abbots Langley
South Road
M SPENCER, Barnoldswick
T FRODSHAM & SON, Macclesfield
DITCHBURN & SON, Peterlee
B EVANS, Blackrock, Co Louth

WHERE TO REPORT FOUND BIRDS
Please notify any changes for inclusion in future issues.

GB The General Manager, Royal Pigeon Racing Association, The Reddings, Nr Cheltenham, Glos GL51 6RN. Phone: 0452 713529.

RSG Ring Manager, The Racing Pigeon Publishing Co Ltd, Unit 13, 21 Wren Street, London WC1X 0HF. Phone: 071-833 5959.

SU The Secretary, Scottish Homing Union, A Shearer, 231A Low Water Rd, Hamilton ML3 7QN. Phone: 0698 286983.

NEHU The Secretary, North of England Homing Union, 58 Ennerdale Road, Walker Dene, Newcastle-upon-Tyne NE6 4DG. Phone: 091-262 5440.

WHU, WALES The Secretary, Welsh Homing Union, T Ash, 3 Coed Cae, Tirfil, New Tredegar, Gwent. Phone: 0443 831161.

IHU 'N' Northern Region Secretary IHU, J Hughes, 34 Adelaide Ave, Belfast BT9 7FY.

IHU 'S' Southern Region Secretary IHU, 69 Lorcan Crescent, Santry, Dublin 9. Phone: 0001-421 391.

NWHU North West Homing Union, P Borrowdale, 77 Hurstbrook, Coppull, Nr Chorley, Lancs. Phone: 0257 793798.

NPA The National Pigeon Association, Mary Roscoe, 32 Derwent Ave, Prescot, Merseyside. Phone: 051-426 2133.

NTU National Tippler Union of Great Britain, B J Rose, 46 Maynard Rd, Hartcliffe, Bristol BS13 0AG. Phone: 0272 647275.

SDU The Secretary, National Union of Short Distance Flyers, 53 Swithenbank Avenue, Ossett, Yorks WF5 9RR.

LPFU The Secretary, Liverpool Pigeon Fliers Union, 3 Wilkin Street, Liverpool 4.

Ringed Wild Pigeons The Secretary, Bird Room, British Museum, London SW7. Phone: 071-636 1555.

BELGE Royale Federation Colombophile Belge, 39 rue de Livourne, Ixelles, 1050—Bruxelles, Belgique. Phone: 010 32 2537 4134.

FRANCE Union des Federations Regionales des Associations des Colombophiles de France, 54 Boulevard Carnot, Lille 59042 (Nord), France. Phone: (20) 068287.

GERMANY (DV) Verband Deutscher Brieftauben Liebhaber, 4300 Essen 1, Postfach 103935, Schönleinstrasse 43. Phone: 010 37 201 778071.

LUXEMBOURG Federation Colombophile Luxembourgeoise, Rue des Champs 16a, Bertrange, Luxembourg.

NORGE Norges Brevdueforbund, Per-Kristian Hansen, Fuglevik Indre 3, 1600 Fredriksdad, Norway.

NL, NPB, NB Bureau NPO, Landjuweel 38, 3905 PH Veenendaal, Holland. Phone: 010 31 838 523 391.

DAN De Danske Brevdueforeninger, Riddersborgvej 9, DK 4900 Naksov, Denmark.

ESP Real Federacion Colombofila Espanola, 34 Eloy Gonzalo, Piso 7°, Madrid, Spain.

PORTUGAL Federacao Portuguesa de Colombofilia, Rua Artur Paiva 30, 1100 Lisboa.

AUSTRIA, ROB, OVB VOB, Waldgasse 788, A-7201 Neudörfl a.d. Leitha. Phone: 010 43 2622 775 925.

SWEDEN M Tage Eriksson, P1 4394 Bjärlöv, 291 90 Kristianstad, Sweden.

CS CSCDR, Maskova 3, Praha 8-Kobylisy 18253, Czechoslovakia.

OLD BIRDS, TRAINING

Ring No	Colour	Sire				

WIDOWHOOD NOTES

DATE PAIRED FOR FIRST ROUND...................... SEPARATED......................

AND RACE POINTS

									Prizes and Positions won

DATE PAIRED FOR SECOND ROUND.................... SEPARATED....................

OLD BIRDS, TRAINING

Ring No	Colour	Sire					

WIDOWHOOD NOTES
DATE PAIRED FOR FIRST ROUND...................... SEPARATED......................

AND RACE POINTS

									Prizes and Positions won

DATE PAIRED FOR SECOND ROUND.................... SEPARATED....................

YOUNG BIRDS, TRAINING

Ring No	Colour	Sire	Dam					

AND RACE POINTS

									Prizes and Positions won

YOUNG BIRDS, TRAINING

Ring No	Colour	Sire	Dam					

AND RACE POINTS

									Prizes and Positions won

STOCK PAIR NO.......................... MATED........................

SIRE   G-SIRE

....

            G-DAM

DAM   G-SIRE

....

            G-DAM

First egg laid	Hatched	Ring No	Colour, Sex. Work done, or how disposed of

Notes and Performances of Sire and Dam:

STOCK PAIR NO........................... MATED.........................

SIRE   G-SIRE

....   {

                              G-DAM

DAM   G-SIRE

....   {

                              G-DAM

First egg laid	Hatched	Ring No	Colour, Sex. Work done, or how disposed of
....	....	....	....
....	....	....	....
....	....	....	....

Notes and Performances of Sire and Dam:

STOCK PAIR NO............................MATED........................

SIRE   G-SIRE

....

           G-DAM

DAM   G-SIRE

....

           G-DAM

First egg laid	Hatched	Ring No	Colour, Sex. Work done, or how disposed of

Notes and Performances of Sire and Dam:

STOCK PAIR NO............................MATED.........................

SIRE   G-SIRE

....

         G-DAM

DAM   G-SIRE

....

         G-DAM

First egg laid	Hatched	Ring No	Colour, Sex. Work done, or how disposed of

Notes and Performances of Sire and Dam:

STOCK PAIR NO............................ MATED............................

SIRE

> G-SIRE

....

> G-DAM

DAM

> G-SIRE

....

> G-DAM

First egg laid	Hatched	Ring No	Colour, Sex. Work done, or how disposed of

Notes and Performances of Sire and Dam:

STOCK PAIR NO...........................**MATED**........................

SIRE   G-SIRE

....

                        G-DAM

DAM   G-SIRE

....

                        G-DAM

First egg laid	Hatched	Ring No	Colour, Sex. Work done, or how disposed of
....	....	....	....
....	....	....	....
....	....	....	....

Notes and Performances of Sire and Dam:

STOCK PAIR NO............................**MATED**...........................

SIRE　　　　　　　　G-SIRE

....

　　　　　　　　　　　G-DAM

DAM　　　　　　　　G-SIRE

....

　　　　　　　　　　　G-DAM

First egg laid	Hatched	Ring No	Colour, Sex. Work done, or how disposed of

Notes and Performances of Sire and Dam:

STOCK PAIR NO............................ **MATED**............................

SIRE	G-SIRE
....	
	G-DAM

DAM	G-SIRE
....	
	G-DAM

First egg laid	Hatched	Ring No	Colour, Sex. Work done, or how disposed of
....			....
....			....
....			....

Notes and Performances of Sire and Dam:

STOCK PAIR NO............................**MATED**..........................

SIRE	G-SIRE
....	
	G-DAM
DAM	G-SIRE
....	
	G-DAM

First egg laid	Hatched	Ring No	Colour, Sex. Work done, or how disposed of
			....
			....
			....

Notes and Performances of Sire and Dam:

STOCK PAIR NO..........................MATED.........................

SIRE	G-SIRE
....	
	G-DAM
DAM	G-SIRE
....	
	G-DAM

First egg laid	Hatched	Ring No	Colour, Sex. Work done, or how disposed of
....	....	....	....
....	....	....	....
....	....	....	....

Notes and Performances of Sire and Dam:

STOCK PAIR NO.......................... MATED..........................

SIRE   G-SIRE

....

          G-DAM

DAM   G-SIRE

....

          G-DAM

First egg laid	Hatched	Ring No	Colour, Sex. Work done, or how disposed of

Notes and Performances of Sire and Dam:

STOCK PAIR NO............................ MATED..........................

SIRE   G-SIRE

....

                              G-DAM

DAM   G-SIRE

....

                              G-DAM

First egg laid	Hatched	Ring No	Colour, Sex. Work done, or how disposed of
....	......	......	....
....	......	......	....
....	......	......	....

Notes and Performances of Sire and Dam:

STOCK PAIR NO............................MATED.........................

SIRE   G-SIRE

....

     G-DAM

DAM   G-SIRE

....

     G-DAM

First egg laid	Hatched	Ring No	Colour, Sex. Work done, or how disposed of

Notes and Performances of Sire and Dam:

STOCK PAIR NO............................MATED........................

SIRE   G-SIRE

..............................{

                              G-DAM

DAM   G-SIRE

..............................{

                              G-DAM

First egg laid	Hatched	Ring No	Colour, Sex. Work done, or how disposed of

Notes and Performances of Sire and Dam:

STOCK PAIR NO............................ **MATED**.........................

SIRE   G-SIRE

....

           G-DAM

DAM   G-SIRE

....

           G-DAM

First egg laid	Hatched	Ring No	Colour, Sex. Work done, or how disposed of

Notes and Performances of Sire and Dam:

STOCK PAIR NO.............................. MATED..............................

SIRE	G-SIRE
.... {	
	G-DAM

DAM	G-SIRE
.... {	
	G-DAM

First egg laid	Hatched	Ring No	Colour, Sex. Work done, or how disposed of
....	....	....	....
....	....	....	....
....	....	....	....

Notes and Performances of Sire and Dam:

STOCK PAIR NO............................. MATED.........................

SIRE   G-SIRE

....

          G-DAM

DAM   G-SIRE

....

          G-DAM

First egg laid	Hatched	Ring No	Colour, Sex. Work done, or how disposed of

Notes and Performances of Sire and Dam:

STOCK PAIR NO............................ MATED..........................

SIRE   G-SIRE

....

                        G-DAM

DAM   G-SIRE

....

                        G-DAM

First egg laid	Hatched	Ring No	Colour, Sex. Work done, or how disposed of

Notes and Performances of Sire and Dam:

RACING PAIR NO.......................... MATED.........................

SIRE... DAM...

Date paired... Separated..

Re-paired... On Widowhood...

First egg laid	Hatched	Ring No	Colour, Sex. Work done, or how disposed of
....	....	....	

RACING PAIR NO.......................... MATED.........................

SIRE... DAM...

Date paired... Separated..

Re-paired... On Widowhood...

First egg laid	Hatched	Ring No	Colour, Sex. Work done, or how disposed of
....	....	....	

RACING PAIR NO........................ MATED........................

SIRE... DAM...

Date paired... Separated...

Re-paired...On Widowhood.....................................

First egg laid	Hatched	Ring No	Colour, Sex. Work done, or how disposed of
....	....	....	....

RACING PAIR NO........................ MATED........................

SIRE... DAM...

Date paired... Separated...

Re-paired...On Widowhood.....................................

First egg laid	Hatched	Ring No	Colour, Sex. Work done, or how disposed of
....	....	....	....

RACING PAIR NO........................... **MATED**...........................

SIRE.. DAM..

Date paired.. Separated..

Re-paired.................................... On Widowhood..

First egg laid	Hatched	Ring No	Colour, Sex. Work done, or how disposed of
....	....	....	

RACING PAIR NO........................... **MATED**...........................

SIRE.. DAM..

Date paired.. Separated..

Re-paired.................................... On Widowhood..

First egg laid	Hatched	Ring No	Colour, Sex. Work done, or how disposed of
....	....	....	

RACING PAIR NO......................... MATED.........................

SIRE... DAM..

Date paired.............................,..................... Separated...

Re-paired.......................................On Widowhood....................................

First egg laid	Hatched	Ring No	Colour, Sex. Work done, or how disposed of

RACING PAIR NO......................... MATED.........................

SIRE... DAM..

Date paired.. Separated...

Re-paired.......................................On Widowhood....................................

First egg laid	Hatched	Ring No	Colour, Sex. Work done, or how disposed of

RACING PAIR NO........................ MATED........................

SIRE.. DAM..

Date paired.. Separated..

Re-paired..On Widowhood..

First egg laid	Hatched	Ring No	Colour, Sex. Work done, or how disposed of

RACING PAIR NO........................ MATED........................

SIRE.. DAM..

Date paired.. Separated..

Re-paired..On Widowhood..

First egg laid	Hatched	Ring No	Colour, Sex. Work done, or how disposed of

RACING PAIR NO........................ MATED........................

SIRE... DAM..

Date paired.. Separated...

Re-paired..On Widowhood.....................................

First egg laid	Hatched	Ring No	Colour, Sex. Work done, or how disposed of

RACING PAIR NO........................ MATED........................

SIRE... DAM..

Date paired.. Separated...

Re-paired..On Widowhood.....................................

First egg laid	Hatched	Ring No	Colour, Sex. Work done, or how disposed of

RACE POINT DISTANCES

Race Point	Code Number	Distance in Miles and Yards

Latitude.................................... Longitude....................................

The Lofts of

A OATES & SON

Owners of the best selection of

ROGER VEREECKE

pigeons in the world,
outside of his own lofts in Deerlijk,
Belgium.

We are the owners of sons and daughters from most
of Roger's best breeders and racers, that have won for
him, top National and International prizes — such as
Young Felix, winner of 1st National Pau 1985; the son
of Young Felix — Kleine Felix, winner of 1st National
Tulle 1990; Para, brother of Young Felix, winner of
5th International Perpignan 1985.

Advertisement continued on next page

A OATES & SON

The Young Felix strain of Roger Vereecke is one of
the most sought after long distance strains in Europe,
having produced outstanding results in National and
International races, not only for Roger Vereecke but
many other top Continental lofts.

Young Felix, one of the world's most outstanding
pigeons. A son of Felix, 1st National Tulle 1978. A
winner of 1st National Pau 1985 and the sire of Kleine
Felix, winner of 1st National Tulle 1990.

Three generations of 1st National winners.

Advertisement continued on next page

A OATES & SON

We will have available a limited number of YBs from our direct Vereecke pigeons, of which we have the best of his best winners bred from winners and breeding winners

Tony Oates with Roger Vereecke seen here inspecting the wing of Young Felix, winner of 1st National and 6th International Pau in 1985.

Advertisement continued on next page

A OATES & SON

Kleine Felix, winner of 1st National Tulle in 1990 for Roger Vereecke. We have in our loft sons and daughters of this remarkable pigeon, the most outstanding son of Young Felix, 1st National Pau and grandson of Felix, 1st National Tulle.

Visitors by appointment only.

YBs from £75 plus carriage.

10% deposit with each order.

Export enquiries welcome.

A OATES & SON

Two Trees, Knott Lane, Rawdon,
Leeds LS19 6JL.
Phone: 0532 503139.

THE MAGER
LONG DISTANCE FAMILY OF PIGEONS
Pau NFC (712 miles) • **Barcelona BICC** (853 miles)

Based upon the Best of British Strains of Ellis, Kirkpatrick, and Newcombe, these pigeons have been consistently timed in from Pau in the Nationals. We have won 2nd Sect K on no less than five occasions and many other section prizes since our first attempt in 1984, these results in the section have brought us over 20 actual Open positions against the best of the National fliers in the country. To further test our pigeons we sent five to Barcelona this year with the BICC, this was a very hard and difficult race for the British contingent, however we timed our first and second pigeons in to win 10th and 17th positions, these being the only two pigeons over 800 miles, with our third and fourth pigeons coming in excellent time behind them for 80% returns. At present we have 24 pigeons in the loft who have flown 712 or 853 miles and been timed at these distances, some as many as four or five times, we keep to the basic good management system but prepare our pigeons thoroughly for the tasks we impose upon them.

Dark chequer cock Bjorn, page 308 October Pictorial, 10th Barcelona 1992, 853 miles.

If you are short on stamina or wish to try your hand at the longer hard races we may be able to help you, we do occasionally have a few birds surplus to nest box accommodation, prices will be quoted on application. *Enquiries to:*

N Mager, High Rising, Town Side, East Halton, Grimsby, South Humberside DN40 3PU. Phone: 0469 40318

Advertisement continued on next page

In their very first race competing at Fed level (SW Glam) against old Widowhood cocks, they won 1st, 3rd, 7th, 8th, 9th, 10th, 11th Open Federation with other fanciers' old birds occupying the other four positions in the first ten.

In the first Fed race for young birds from Clee Hill (1,241 birds), eight dropped together to win 1st, 2nd, 3rd, 4th, 9th, 16th, 17th, 19th SW Glam Fed.

In the first Welsh North Road race on the same day my other team won 1st, 2nd, 3rd, 4th, 5th, 6th, 7th Club and 1st, 2nd, 3rd, 5th, 6th, 7th, 8th Sect (1,240 birds).

The following week Clee Hill again, I won 1st, 2nd, 8th SW Glam Fed (2,108 birds). This was a very hard race, vel 977. In the same race WNR I finished 8th Sect (2,660 birds).

From Shrewsbury SW Glam Fed I won 4th, 9th, 14th Fed (2,193 birds), and was only 1.4 ypm behind the Fed winner.

In the WNR from Shrewsbury I won on August 2, 1st, 2nd, 3rd Club, etc, also 2nd, 11th, 12th Sect (3,062 birds).

From Nantwich I won 1st, 9th Open SW Glam Fed (1,872 birds). Then from Beeston WNR finished 1st, 2nd, 3rd, 4th Club, 2nd, 9th, 10th, 12th Sect (2,415 birds). My birds continued in this winning vein right through to the end of the season and topped the SW Glam Fed in the last two races from Clee Hill and Carlisle, ie Clee Hill (comeback) 1st, 3rd, 7th, 9th, 10th, 14th, 15th Fed (1,200 birds); Carlisle, 1st, 2nd, 12th, 14th, 19th SW Glam Fed (9,641 birds). In the previous Carlisle race I clocked six birds in the WG National before any other fancier clocked six, winning £750 in the National.

In the very last young bird race in Llantwit Fardre Club I clocked seven birds to finish probable 1st, 2nd, 3rd, 4th, 5th, 7th, 8th Club. This race was very hard indeed, the birds hit heavy rain five minutes after liberating and had 50 miles of rain in front of them. However I was half-hour ahead of other members of the club and am likely to have seven birds in the first 30 in the WNR Sect results. Full result not yet to hand.

These birds are winning for fanciers the length and breadth of the country but this season pride of place must go out to Ron Dudley of Abergavenny on winning the Welsh South Road Grand National from Ventnor. The dam was from my No 1 pair and I selected a cock for Ron to pair the hen to from the Moris/Scheers sale in 1991. The following week Ron topped the Rhondda Valley Fed from Reading with the nestmate.

Young birds for sale in 1993, £35 each or £200 for six from children of direct imports, or £55 each or £300 for six from direct imports and proven breeders of winners.

For all fanciers who order six young birds in 1993 I will give my feeding system to (there are different products in the water every day). This has proven to be successful to fanciers such as Cary Grother, runner-up to me in the Fed Championship 1992; Bob Garn of Bridgend, winner of fifteen 1sts 1992, many of these with my birds; Peter Owtrim of Llantwit Fardre, again who has often won the first four positions in club, etc.

For sale and pedigree details apply in confidence to:

JOHN DAVID
1 Llest Cotts, Llantwit Fardre
Nr Pontypridd, Mid Glam, S Wales
Phone: (0443) 202732

Late news: *I am winner of YB and Combined averages in Llantwit Fardre Club 1992, breaking the all-time points record for YBs since the club was formed.*

GABY VANDENABEELE

HOME ALONE
GB91P54218
1992 RESULTS

1st	CHELTENHAM	2439P
1st	MANGOTSFIELD	2295P
1st	GLOUCESTER	2174P
1st	CHELTENHAM	1237P
1st	MANGOTSFIELD	460P
1st	GLOUCESTER	430P
1st	MANGOTSFIELD	388P
1st	CHELTENHAM	239P
1st	MANGOTSFIELD 2/B	72P

DAVE HOLDERNESS
PRESTON

Frank Laird
ST HELENS

DE VADDER SPRINTERS

Champion Crapul, winner of sixty-two prizes. In his first year of breeding for us he bred four youngsters all of which won. He is typical of the De Vadder pigeons.

We have at stock 20 pairs of De Vadder sprinters, which were purchased direct from the De Vadder lofts, from all of their best stock pigeons. A few of De Vadder's results (summary of the last 18 years): 9,800 prizes of which 629 were pure 1sts, an average of fifty-four 1sts per year. General National Champion of the Regie der Posterijen and the RTT General All-Round Champion of the City of Groot — Haaltert with 15 honourable mentions 1st, 2nd, 4th, 5th, 6th & 9th in the Provincial Championship of the KBDB division Dender-Vallei — Ninove — Half-fond — Young Ones.

Since we introduced the De Vadders to our racing team we found them very fast pigeons which have won for us against the odds. If you are looking for something to add speed to your birds these would make an ideal cross or found a family of sprinters.

Young birds £40 each, £200 for six plus £12 carriage.

Mr & Mrs P G Clare
Cherrytree Farm, Hobhole Bank, Old Leake, Boston, Lincs PE22 9RT.
Phone: 0205 750295 or 0205 351773.

"Having been a personal friend of the De Vadder family for many years, I purchased for Geoff and Bridget a complete round of youngsters from their best breeders and racers. They have matured into superb pigeons and are a credit to such good fanciers".—**Frank Tasker**

ROGER SUTTON
OF CONGLETON

Van der Weyers — D Mitcham and Direct (Sprint)
Andre Berte Janssens (Middle Distance)
Delbars and Krauths (Long Distance)

Representatives of the above families are usually for sale.
The parents of these birds have won multiple 1st prizes or are brothers and sisters of big race winners. These birds are well developed and will make excellent stock pigeons.

Prices from £25. There will also be some young birds available during 1993.

Recent results:

1988 — Thirty 1st prizes including 1st Club, 2nd Fed Bath 2,612 birds; 1st Club, 2nd Fed Newton Abbot 2,286 birds; 1st Club, 19th Fed Weymouth 6,615 birds; 1st Club, 1st Fed Sartilly 3,412 birds; 1st Club, 4th Fed Rennes 1,067 birds; 1st Club, 1st Fed Nantes 1,669 birds; 1st Lancs Social Circle Nantes; 1st Club, 7th Fed Saintes 1,147 birds; 1st Club, 5th Fed Niort; 3rd Great Northern Saintes; 1st Club, 6th Fed Bath 2,261 birds; 1st Club, 2nd Fed Bath 2,318 birds.

1989 — twenty-eight 1st prizes including 1st Club, 8th Fed Gloucester; 1st Club, 15th Fed Bath; 1st Club, 13th Fed Bath 7,529 birds; 1st Club, 1st Fed Wincanton; 1st Club, 5th Fed Wincanton; 1st Club, 2nd Fed Weymouth 7,043 birds; 1st Club, 3rd Fed Wincanton; 7th & 10th Sect L, 313th & 337th Open NFC Nantes; 1st Club, 5th Fed Rennes; 1st Club, 4th Fed Weymouth; 1st Club, 3rd Fed Rennes; 2nd Cheshire 2-Bird Nantes; 1st Lancs Social Circle Niort; 6th Great Northern Saintes; 1st Club, 1st Fed Gloucester; 1st Club, 8th Fed Bath; 1st, 9th & 11th Sect L NFC Cherbourg; 1st Cheshire 2-Bird Bridgnorth.

1990 — twenty-eight 1st prizes including 1st Club, 5th Fed Frome; 1st Club, 2nd Fed Frome; 1st Club, 8th Fed Wincanton 7,518 birds; 1st Club, 1st Fed Weymouth; 1st Club, 4th Fed Rennes 3,250 birds; 1st Club, 11th Fed Dorchester; 4th Sect L, 443rd Open NFC Nantes; 1st Club, 7th Fed Rennes; 1st Club, 1st Fed Wincanton; 10th, 12th & 13th Great Northern Angouleme (sent three — only 17 on the day); 1st Club, 14th Fed Bath 6,587 birds; 1st Club, 6th Fed Frome; 1st Lancs Social Circle Weymouth.

1991 — twenty-one 1st prizes including 1st Club, 1st Fed Gloucester 5,953 birds; 1st Club, 25th Fed Mangotsfield 7,204 birds; 1st Club, 1st Fed Hullavington; 1st Club, 3rd Fed Dorchester; 1st Club, 10th Fed Wincanton 5,327 birds; 1st Club, 6th Fed, 15th Combine Rennes; 12th & 39th Sect L, 63rd & 366th Open NFC Nantes; 1st Club, 3rd Fed Weymouth; 1st Club, 1st Fed Nantes; 16th & 46th Sect L, 135th & 440th Open NFC YB St Malo; 2nd Middlewich Open Sartilly (Champion of Cheshire).

1992 (Old Birds only) — Twenty 1st prizes including 1st Club, 3rd Fed Hullavington; 1st & 2nd Club, 1st & 3rd Fed Wincanton; 1st Club, 2nd Fed Dorchester; 1st Club, 12th Fed Weymouth 7,505 birds; 1st, 4th & 12th Bass Yearling Classic Rennes; 4th Sect, 21st Open Mid Nat Nantes; 1st Lancs Social Circle Nantes; 1st Club, 2nd Fed Niort; 1st Keele 2-Bird Niort; 1st Club, 22nd Fed Angouleme. Winning ten different Channel races.

Visitors are always welcome — please telephone to check availability and to arrange times.

J R Sutton, Bent Farm, Astbury, Congleton, Cheshire CW12 4RW. Phone: Congleton (0260) 273677.

WOW!
QUALITY! QUALITY!
If anyone can help Doddy can!

On offer will be youngsters from my two families, my greater distance strain and my Busschaerts.

The greater distance strain is based on the **Kirkpatricks** from **R Thompson of Annan** and direct children of the **Berwick** hen. Other introductions are from such pigeons as **Knightsdale Lady** 1st Open Palamos; **Ramona** 1st British National Barcelona, 1st Sect, 2nd Open Palamos 700 miles; **Woodsider** 1st Open Palamos 800 miles; **Palamos Pixie** 1st Open Palamos; **Big Fella** 1st and 2nd Open Fraserburgh 425 miles; **Versatile** 1st Sect, 3rd Open Palamos; **Serene** 1st Sect, 36th Open Palamos 810 miles.

Youngsters will be available from sons and daughters of Palamos Supreme, Fairoak Supreme, Fairoak Leader, Palamos Pixie, Mike Young's Channel Queen, R & M Venner's double Lerwick winner.

The price for these, **£50 each.** From Fairoak Leader **£100 each.**

BUSSCHAERTS

From the great men direct Billy Parkes and Dr J J Horn. From such pigeons as the **Red Unrung,** son of the **Henson Hen,** daughter **Barney x Smart Bird,** daughter of **Mr Busschaert x Palmer Hen,** two daughters of **Charter Flight,** son and daughter off **The Wilkinson Cock,** son and daughter inbred to the **Newton Pair,** also others such as son of **Studtopper x Bess,** daughter **Clapper x Bess,** the **Staddon Hen** a great producer son of **Coppi 046,** daughter of **Starview Pegasus,** daughter of **Starview Leo.**

Price of these youngsters from stock birds from **£50 to £100.** Off of race birds **£30.**

Advertisement continued on next page

Continued from previous page

If you are interested in prize cards only or pool money or
the odd TV set or a clock.

The answer is here.

Prize cards, just a few: 1st Palamos 694 miles; 1st
Perpignan BICC 634 miles; 1st Pau BICC 544 miles; Pau
2nd, 12th, 20th, 27th and 16th Sect, 57th, 145th, 54th, 66th
and 129th Open; £1,000 for 1st Rainey Saints, 2nd, 2nd,
2nd and 6th Sect, 4th, 13th, 20th, 5th, 5th, 9th and 11th
Open. I could go on and on, money won runs into thousands
of pounds.

Note: How these pigeons do for others. I gave seven young
birds to Ken Otton this year, Doddy's Gift won 1st Open,
1st Sect British Barcelona Club; Doddy's June won 1st West
of England Combine Littlehampton 2,640 birds; Doddy's
White Flight won 37th Open 3,611 birds, 15th Sect 1,017
birds Guernsey Classic Flying Club. Other fanciers who
have won with these pigeons are: K & H Jones, Taunton,
1st & 2nd Combine approx 7,000 birds; D Bridges, Glos,
1st Combine approx 6,000 birds; A Thompson, Dunstable,
1st Combine; P Proctor, Newbury, 1st Combine. In 1990
West of England SR Combine, five pigeons in the first
seven, 7,805 birds for four different fanciers.

Don't delay book your youngsters today.

I will have a few old birds for sale, prices on request.

Order from:

TREVOR DODD
"Fair Oaks", Ham,
Creech St Michael,
Nr Taunton, Somerset TA3 5NY.
Phone: 0823 442155.

BRAMBLE PARK
LOFTS

The home of probably one of the finest collections of winning

JANSSENS
in existence today.

This has been achieved firstly by only breeding from performance pigeons, that have been carefully selected vigour, type, etc, which in turn are capable of passing these qualities onto their offspring.

B C ML82.297316, son of Superman, stock cock Janssen.

1st Sect, 1st Open, 3,299 birds Hexham, Notts & Dist NR Fed. Janssen cock.

Bramble Park Lofts own many top Janssen bloodlines including a son of Superman, thirteen 1sts, now owned by Planet Bros. Grandchildren of Oude Merckx, Bange of 59, Witoger of 65, Oude Geeloger, plus the best Meulemans, A Wouters and Van den Bosch via Albert Babbington.

These Janssens, along with Van Wildemeersch, have enabled me to win two hundred and thirty-six 1sts, one hundred and eighty-six 2nds, one hundred and eight 3rds. Twenty-seven 1sts 1988. Twenty-six 1sts 1989. Top prize-winners year after year.

Advertisement continued on next page

Continued from previous page

Also at Bramble Park Lofts are housed my second family of pigeons. These are the very good

VAN WILDEMEERSCH

Once again with careful selection I now own a family that can win right through to Thurso nearly 400 miles.

My Wildemeersch include birds bred down from Trimard and The Lehore, The Black Cock and Feathers. The Krop and Princess, Orleanske, Son of Merckx of 72, Klepperke, Wittepen Kourre. A double grandson of Gouden Beer and daughter of Bonten and Sels Duiven. Vuile Gaby, Kleine Bliksem, Zorro, Rene, Loper, Wittekele, Tenke.

1st Open Northallerton, 5,625 birds. Notts & Dist NR Fed. Wildemeersch.

85S65196. Double grandson of the great Orleanskie. Van Wildemeersch.

End of 1989 season I decided not to participate in any more racing due to old age and illness, hence all my youngsters will be for sale.

All one price — £30 each or £150 for six.

Please add £10 for carriage.

A full pedigree supplied with each youngster.

GEORGE CROWDER

57 6th Avenue, Clipstone, Nr Mansfield, Notts NG21 9DN

Phone: 0623 28094

FRANS VAN WILDEMEERSCH

The very best bloodlines of this family are housed at Hill View Farm including the famous Blue Beer, 79.3470660, one of the very best ever. A fantastic producer like his sire. Also Laaste Beer, sire to many winners.

Plus eight other sons and daughters of the famous **Beer** of **73.3431965.**

There are nine direct children from **Bliksem** and **Kleine Bliksem,** Frans' present day champion. One son breeding two Fed winners in one nest.

In fact, I have bred several nests from the Van Wildemeersch pigeons with two winners in one nest. I also have hens to match these class cocks such as **The 503,** mate of the **Blue Beer** and her daughter **Pied Hen** who have bred scores of 1st prize-winners.

We have children of **Orleanske, Pronostiek Duivin, Ace Producer, The Roger, 3rd National** for Frans.

No 1 Cock, Blue Beer,
79.3470660.

Son of Orleanske,
81.3300172.

Advertisement continued on next page

Continued from previous page

Daughter of Beer, crack breeder, sister to The Hundred, 79.3233518.

Son of Parel, 80.3001240.

Son of Beer and Pronostiek Duivin, GB83L65122.

Producer Hen, GB80Z12031.

Son of Bliksem, 82.3430727.

No 3 Cock, Laaste Beer, 81.3258990.

Son of Jonge Bliksem, 86.3006156.

Advertisement continued on next page

Continued from previous page

**Het Talent, daughter of
Beer, 81.3300162.**

**Daughter of Bliksem,
82.3430800.**

**Son of Prins, ace
producer, 80.3493365.**

**Champion, 503 Hen,
daughter of Merckx and
Pronostiek Duivin.**

Plus many more of this calibre, too many to list.

I have stock for sale from **£30.** You could pay much more
elsewhere but could not buy better.

MR & MRS LEVITT
**Hill View Farm, Wilsic Lane, Tickhill,
Doncaster, S Yorks DN11 9LI.**
Phone: Doncaster 745939

BILL JOHNSON BUSSCHAERTS

MR LONDON NORTH ROAD COMBINE

*Without doubt the top Busschaert flier over the long distance is "King Bill" of Thornton Heath, better known as "Mr London NR Combine". These accolades are laid firmly at the feet of Bill Johnson of Thornton Heath who, in my opinion, has proved the country's No 1 long distance Busschaert flier. His family, based on the Old Man is a legend. Performances of Old Man's progeny include a g.son 1st Up North Combine Bourges 675 miles for C Copeland. G.son 2nd Open LNRC Morpeth for J Heywood. A son was 25th Open LNRC Thurso for me, these in 1987. In 1988 dtr were 3rd Open LNRC Stonehaven, 12th Open LNRC Thurso and a g.dtr 25th Open LNRC for me. A g.dtr was 3rd Open NFC YB for J Norris. There has never been another fancier to compare with Bill Johnson as a LNRC fancier during the past 12 years. He has dominated the Combine in my view. For my money, in the Combine style of racing, Bill Johnson is the No 1 competitor in the country.—***F G Wilson.***

*In my opinion Bill Johnson rates as the hottest thing to hit the London racing scene since the Alf Baker purple period. Above all it must be remembered that what Bill has achieved with his pigeons has been all against the grain. Despite his unfavourable loft location Bill has been riding high for over 12 years now and during this spell he has amassed a string of prizes no other Combine member can hold a candle to. One of the most important things to take into consideration when it comes to selecting a would-be vendor for your stock purchases is whether or not the performances have been put up by a team of birds rather than just one or two loft stars. Bill's birds undoubtedly fly as a team and queue up to go into the clock on race days as any LNRC Catford station clocksetter will vouch for.—***Doug Went (The Megaphone).***

THE NO 1 LOFT LNRC

I can assure any would-be purchaser they are without equal. They give pleasure and in these expensive times they more than pay for themselves. If you want G Busschaerts to fly the distances from 90 miles to 600 miles they will and in the London area have no equals. Consistency and team results are what we all try to achieve. This team of Busschaerts has done just that. These direct Busschaerts on offer were all hand picked by me for their Klaren/Sooten breeding, The Old Man's bloodlines, ask Tom Hoadley and see The RP 28 December 1984 and others.

These Busschaerts of mine are winning up and down the country. All race results are mine, I don't need to sell my birds on other people's results. Fuller pedigrees and information to purchasers. *See Pictorial No 179 Busschaert Special.* My results in LNRC have been achieved against the cream of the Fancy and when in my opinion any fancier intending to introduce new stock or anyone just starting, one should ask one's self, what class of competition this fellow flies in? LNRC is the largest Combine flying North or South. In the whole of the South of England LNRF is the largest in the LNRC and possibly the largest in the South East. Thornton Heath RPC is one of the strongest clubs in the LNRF.

LNRC RESULTS 1972 TO 1988

LNRC Berwick (311 miles). Open results (av 9,000 birds): twice 2nd, 3rd, six times 5th, 7th, twice 8th, 9th, twice 12th, twice 13th, twice 14th, twice 15th, 16th, 17th, 22nd, 23rd, 24th, 57th & 59th Open, 11th & 12th Sect, plus 26 other positions in the

Advertisement continued on next page

PIGEON PHOTOGRAPHY BY PETER BENNETT

MYRTLE LOFTS

Based on the two essentials — pedigree and performance

GENUINE

KAREL HERMANS

Our birds originated from Karel Hermans (Belgium) prior to his death. We have children and grandchildren from his following champions: Meil, Jef, Two Blacks, The Velthem, Scherpe, Bordeaux, The Hourban, Westerloo, *plus many others.* I believe we were the only loft in the country to have The Westerloo and Hourban pigeons. We blended many champions into these to establish an outstanding stock team of Hermans, such as brothers and sisters to the outstanding Hermans hen, 1,001, a winner of thirty 1sts, plus three 1sts Fed; Treble 8, an outstanding Fed Topper with many other prizes, as a YB this hen bred four Fed toppers in one season, all sold to our regular customers; Double 8, Fed topper with eight 1sts in two seasons; Dawn Run, a winner of 30 prize cards, thirteen 1sts, three 1sts Fed, 14 Fed cards in first ten. The old Belgian cock known as Double 2, sire and g.sire to over 100 prizewinners, ex: Champion Quartz with eighteen 1sts, now at Louella Stud; another of his unraced daughters sold for £520 in auction; Hopalong, a good Fed topper with eleven 1sts; the Whitley Hen, bred by us, dam of sixty 1st prizewinners; Money Spinner, a winner to date for us of £1,402 — we could go on.

The Hermans have made us top prizewinners on 19 occasions, in clubs up to 80 members. Seven breeder/buyer awards.

NO MOB FLYING — NO LOFT ADVANTAGE.

1992 results to date with six weeks left to race — TWELVE 1sts, with 59 prize cards, 14 Fed positions.

BLUE GEM	**MONEY SPINNER**	**DAWN RUN**
Dam to endless winners. A winner of eleven 1sts, three 1st Fed, YB National, Bird of the Year award (250 miles), winner of OB National, Bird of the Year award. We have brothers and sisters, children and grandchildren in our stock lofts breeding winners. The best of champion Bluebell blood. Raced by Firman & Dorman.	Proven dam. This hen is a winner for us of £1,402, winning from 100 to 400 miles. She has been very lightly raced and lightly bred from, nevertheless bred winners to two different cocks. Each year we save this hen to race in the Classics where she excels. This hen will be an asset to our stock loft when retired.	Dam to endless winners. A winner of 30 prize cards. Thirteen 1sts, three 1sts Fed, 14 Fed cards in first ten. We have many children and grandchildren breeding winners in our stock lofts.

Our birds don't only win for us, they win for the many fanciers who come back to us year after year, which speaks volumes.

Advertisement continued on next page

SEA VIEW LOFTS
1967-1992
1st Up North Combine Melun, 10,920 birds, vel 1249.5, 1991

Our family of birds are based upon **Vandeveldes, Hermans, Bob Tinkler** and the late **Tom Kilner,** blended together over the years to make an ideal family of pigeons for racing in middle to long distance events.

In 1984 we introduced birds direct from **William Geerts,** Schilde, Antwerp, Belgium. They were an instant success winning at all distances from Selby 68 miles to Bourges 557 miles, winning club, Fed, Sect, Open races and championship clubs.

In 1991 we decided to look for another cross, so we went to **Frank Aarts,** Tilburg, Holland, and introduced his best birds. These are **Janssen, Janssen Vandenbosch and Horeman.** Frank also used the **William Geerts** birds for his family. Our birds carry the bloodlines of De Rode Van Gompel, Bange, Taxi (W Geerts), Blauw Pieterse, Schallie Pieterse, 100% Janssen, Witte Kas, Kneet, Wesel, Het-Tam, all champion breeders for Frank.

Storm Queen, Geerts. Winner of 5th UNC Lillers, 331 miles, 13,732 birds; 1st TRCC, 1st NECC, £700. Dam and grand-dam of winners etc.

Dark cock, Geerts. Winner of 76th UNC Eastbourne, 19,580 birds; 1st Two Rivers CC, £400; also 36th UNC Lillers, 12,000 birds 1992. Son of Storm Queen.

Geert. Winner of 13th UNC Lillers, 13,714 birds; 9th UNC Lillers, 13,732 birds; 186th UNC Abbeville, 19,061 birds.

Broken Legs, Geerts x Vandevelde. Winner of 7th UNC Clermont, 7,251 birds, £800 plus video. Dam of Broken Legs won £1,263.

Blue cock, direct William Geerts. Sire of Storm Queen, 5th UNC and Geert, 9th, 13th and 186th UNC. Also Broken Legs, 7th UNC Clermont and dark cock, 36th UNC Lillers, 76th Eastbourne. Also Fed, Sect, Championship Club winners.

Young Tommy. Winner of Bourges, 555 miles, five times 1985, 1986, 1987, 1988 & 1989, winning 33rd, 51st, 82nd & 141st Up North Combine. Also 53rd UNC Oudenaarde 1985, 11,603 birds. Sire and grandsire of winners.

We will have youngsters for sale during 1993 season from £30 to £50 each.
Satisfaction assured at all times.
Visitors welcome, but please phone first.
Apply in confident to:

F BASLEY & G ROBERTS
33 Ninth St, Horden, Peterlee, Co Durham SR8 4LZ. Phone: 091-586 9068

GREEN LANE LOFTS

Original Home of the Ebony Krauths

Champion Flying Ebony, described as the world's No 1 black Krauth cock was bred and raced at the Green Lane Lofts. There are eight direct sons and daughters at stock.

These are the original old Krauths from such Krauth champions as The Flying Dutchman, 1st International Barcelona (746 miles), Vierzon, three 1sts at 600 miles and 1st Dax (650 miles); and the Oude Chek Cock, 18 times a prize-winner from 600 to 746 miles.

BLACK KING   **BLACK BULL**   **BLACK DREAM**

With birds from these lofts, fanciers have won major races in England, Scotland, Ireland and Wales including:

- 1st West Sect Scottish National Sartilly (498 miles);
- 1st West Sect Scottish National Rennes 1991 (540 miles);
- 1st Welsh NR Fed Burscough (21,000 birds);
- 1st & 2nd Club, 1st & 2nd Fed, 1st & 2nd Amal, 13th & 39th Open Irish Derby Nantes (470 miles — no day birds), plus £1,026;
- Three times 1st Sect NIPA (8-9,000 birds);
- First 15 in South Down Fed etc.

Advertisement continued on next page

Firsts have been won in the following Feds, Amals, Combines through to 515 miles and up to nineteen 1sts:

Warrington Fed, Wigan Amal, Wigan Fed, New Lancs Fed, West Coast Fed, Liverpool Amal, Pentland Fed, South Lanarkshire Fed, Berks & Bucks Fed, Welsh NR Fed, Devon Fed, Cardiff Fed, Altrincham Fed, South Down Fed, County Down Amal, London NR Fed, East Anglian Fed, East Anglian Classic, South Lancs Combine, Midland Counties Combine, Four Counties Combine, Lancashire Combine.

In my last season racing I won twenty-two 1sts, with five 1sts Fed, including the first seven positions in Fed. The last eight times competing at Niort (504 miles) won six 1sts and twice 2nd. In last Niort race six birds sent and five clocked on day winning 1st, with three of those birds having flown Nantes (434 miles) on the day the previous Saturday, again winning 1st (840 miles in a week!).

★   ★   ★   ★   ★   ★

I offer some superb youngsters for sale.

A few older birds for sale occasionally.

Satisfaction is assured.

★   ★   ★   ★   ★   ★

HAROLD HART

232 Green Lane, Leigh, Lancashire WN7 2TW.
Phone: 0942 606030

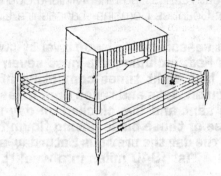

Advertisement continued on next page

L TUPLIN & SON

GALA CHAMPION 04
1st Sect Fed 1,210 birds, 1st Open Huntingdon 5,352 birds; 1st Sect Fed 682 birds, 1st Open Eastbourne 3,006 birds; 1st Sect Fed 867 birds, 1st Open Tonbridge 2,853 birds; 2nd Sect Fed 1,079 birds, 2nd Open Huntingdon 4,719 birds; 3rd Sect Fed 937 birds, 3rd Open Tonbridge 3,477 birds; 4th Sect Fed 902 birds, 7th Open Peterborough 3,982 birds; 6th Sect Fed 990 birds, 8th Open Eastbourne 4,336 birds; 5th Sect Fed 806 birds, 11th Open Eastbourne 3,199 birds; 6th Sect Fed 1,139 birds, 12th Open Eastbourne 4,604 birds; 6th Sect Fed 850 birds, 14th Open Tonbridge 3,357 birds; 4th Sect Fed 986 birds, 15th Open Huntingdon 4,798 birds.

GALA CHAMPION 43
1st Sect Fed 893 birds, 1st Open Eastbourne 4,450 birds; 1st Sect Fed 805 birds, 1st Open Eastbourne 3,928 birds; 1st Sect Fed 850 birds, 1st Open Tonbridge 3,357 birds; 2nd Sect Fed 1,059 birds, 3rd Open Romford 4,862 birds; 2nd Sect Fed 903 birds, 4th Open Eastbourne 4,169 birds; 3rd Sect Fed 764 birds, 4th Open Tonbridge 3,011 birds; 4th Sect Fed 682 birds, 4th Open Eastbourne 3,006 birds; 5th Sect Fed 798 birds, 12th Open Eastbourne 3,422 birds; 9th Sect Fed 845 birds, 16th Open Stevenage 4,341 birds; 10th Sect Fed 717 birds, 18th Open Eastbourne 3,001 birds.

GALA POOLER
1st Sect Fed 1,079 birds, 1st Open Huntingdon 4,719 birds; 1st Sect Fed 902 birds, 1st Open Peterborough 3,982 birds; 2nd Sect Fed 806 birds, 2nd Open Eastbourne 3,199 birds; 3rd Sect Fed 1,210 birds, 3rd Open Huntingdon 5,352 birds; 3rd Sect Fed 1,266 birds, 3rd Open Stevenage 4,867 birds; 4th Sect Fed 952 birds, 4th Open Peterborough 3,882 birds; 4th Sect Fed 1,059 birds, 5th Open Romford 4,862 birds; 5th Sect Fed 867 birds, 5th Open Tonbridge 2,853 birds; 4th Sect Fed 990 birds, 6th Open Eastbourne 4,336 birds; 8th Sect Fed 903 birds, 14th Open Eastbourne 4,169 birds; 6th Sect Fed 1,025 birds, 17th Open Eastbourne 4,466 birds.

GALA RED ARROW
1st Sect Fed 1,266 birds, 1st Open Stevenage 4,867 birds; 1st Sect Fed 1,215 birds, 1st Open Tonbridge 4,969 birds; 1st Sect Fed 764 birds, 2nd Open Tonbridge 3,011 birds; 2nd Sect Fed 798 birds, 3rd Open Eastbourne 3,422 birds; 6th Sect Fed Tonbridge 1,196 birds; 8th Sect Fed Ashford 1,154 birds.

Advertisement continued on next page

Continued from previous page

L TUPLIN & SON

GALA MOURNER
1st Sect Fed 205 birds, 3rd Open Nevers 1,221 birds; 5th Sect Fed 1,139 birds, 8th Open Eastbourne 4,604 birds. Sire of Gala Mourner is Gala Champion Producer.

GALA TV
1st Sect Fed 368 birds, 3rd Open Clermont 1,533 birds; 4th Sect Fed 285 birds, 8th Open Melun 1,319 birds; 1st Sect Fed 479 birds, 8th Open Clermont 2,119 birds; 1st Lincs 2-BC Melun; 1st Lincs 2-BC Clermont. Gala TV is a grandson of Gala Champion 04.

GALA MEMORY
1st Sect Fed 376 birds, 1st Open Clermont 1,751 birds.

GALA LADY
1st Sect Fed 322 birds, 10th Open Le Mans 1,525 birds; 3rd Sect Fed 140 birds, 12th Open Niort 718 birds; 6th Sect Fed 590 birds Eastbourne; 11th Sect Fed 205 birds Nevers; 1st Lincs 2-BC Le Mans; 2nd Lincs 2-BC Niort

Advertisement continued on next page

Continued from previous page

L TUPLIN & SON

GALA CRACK

1st Sect Fed 806 birds, 1st Open Eastbourne 3,199 birds; 2nd Sect Fed 1,266 birds, 2nd Open Stevenage 4,867 birds; 4th Sect Fed 1,215 birds, 6th Open Tonbridge 4,969 birds; 4th Sect Fed 776 birds, 6th Open Eastbourne 3,121 birds; 4th Sect Fed 1,139 birds, 7th Open Eastbourne 4,604 birds; 12th Sect Fed 952 birds, 15th Open Peterborough 3,882 birds. Gala Crack is son of Gala Pied Man.

GALA ACE

1st Sect Fed 776 birds, 1st Open Eastbourne 3,121 birds; 2nd Sect Fed 1,215 birds, 2nd Open Tonbridge 4,969 birds; 2nd Sect Fed 902 birds, 2nd Open Peterborough 3,982 birds; 2nd Sect Fed 937 birds, 2nd Open Tonbridge 3,477 birds; 1st Sect Fed 798 birds, 2nd Open Eastbourne 3,422 birds; 2nd Sect Fed 867 birds, 2nd Open Tonbridge 2,853 birds; 2nd Sect Fed 1,154 birds, 3rd Open Ashford 4,810 birds; 5th Sect Fed 764 birds, 6th Open Tonbridge 3,011 birds; 5th Sect Fed 990 birds, 7th Open Eastbourne 4,336 birds; 3rd Sect Fed 850 birds, 7th Open Tonbridge 3,357 birds.

GALA PIED MAN

Champion stock cock. Sire of many winners including Gala Crack.

GALA 62

2nd Sect Fed 1,120 birds, 2nd Open Tonbridge 5,108 birds; 2nd Sect Fed 805 birds, 2nd Open Eastbourne 3,928 birds; 1st Sect Fed 717 birds, 2nd Open Eastbourne 3,001 birds; 3rd Sect Fed 899 birds, 3rd Open Romford 4,746 birds; 2nd Sect Fed 818 birds, 4th Open Tonbridge 3,123 birds; 3rd Sect Fed 637 birds, 10th Open Tonbridge 2,976 birds; 7th Sect Fed 656 birds, 14th Open Tonbridge 2,830 birds; 6th Sect Fed 850 birds, 15th Open Tonbridge 3,357 birds; 5th Sect Fed 870 birds, 20th Open Eastbourne 4,398 birds.

Advertisement continued on next page

Continued from previous page

L TUPLIN & SON

GALA LEADER
1st Open Midland National Flying Club Rennes. Son of Gala Prince.

GALA PRINCE
1st Open Melun 1,569 birds beating second bird by over 25 yards, only five birds in Fed on day; 1st Sect Fed, 1st Open Melun 1,480 birds; 1st Lincs 2-BC Melun by over 34 yards.

GALA BLUE GIRL
11th Fed Clermont; 7th Fed Clermont 2,419 birds. Dam and grand-dam of winners. One of the top breeding hens.

GALA CHAMPION PRODUCER
2nd Sect Fed 428 birds, 14th Open Falaise 2,107 birds. Gala Champion Producer is a top class stock cock of Channel winners, sire of Gala Mourner.

THERE IS NO SUBSTITUTE FOR QUALITY

1993 youngsters available from £40 each plus carriage. **From our champion stock birds and top winners.** *Prices on application.*
£10 deposit per youngster when ordering. *Satisfaction assured.* None sold locally. Visitors strictly by appointment only. SAE please with all enquiries.

L TUPLIN & SON, Galabank Lofts
Park View, Legsby, Market Rasen, Lincolnshire
LN8 3QW. Phone: Market Rasen (0673) 843272

CHURCH VIEW LOFTS

Every family housed at Church View Lofts comes from elite lofts flying to top racing standards *today — not years ago.*

Every family housed at Church View Lofts is producing birds that win *under UK conditions.*

We are dedicated to quality and to results — not numbers.

Quality in the performance background of our stock birds.

Quality in their environment.

Quality in the health and condition of our YBs.

SPRINTERS

MAURICE VERHEYE — The bloodlines of Young Brives, Groten Brives, Frankie and Dark Merckx are producing Fed winning stars year after year. These are some of the top Verheyes in the UK.

MARCEL DEMEESTER — The fantastic scoring record of The Sprintmaster is already being repeated here at all velocities. We have made major new additions for 1993 because of this. Reet eaters!

RAES & SON — Sprint experts from Ghent, 14 times Combine champions. Many wins reported in the UK from these striking pigeons.

SPRINT/MIDDLE

MICHEL DELEN — The beautiful all-weather sprinters of this Antwerp Union star are now here. Jacobs/Geerts/Meuleman-based with the blood of De Buffel, Boogaart, Witstaart and Straaljager.

ROMAIN LEGIEST JANSSEN — Looks and ability from the Gold Award winner. Super Janssens breeding Fed winners in the UK.

ALBERT VAN MIERT JANSSEN — Pure Janssens of Arendonk from the old master at Turnhout. The perfect blend with our Legiest blood.

DESMET BROS — One of the very top lofts in Belgium, winning a staggering thirty-two 1sts in 1992. Winning from 50 to 500 miles. We have a small but high scoring stud of Andre Desmet's best.

MIDDLE & LONG DISTANCE

SILVERE TOYE — We have proved beyond doubt the strength of the Toye originals here with the blood of Tarzan, Dikkop and Peter Pau. For 1993 we introduce direct birds from all the new champions, Darco, Youngblood, Dali and Aramis. The unequalled Church View Toyes.

ALFONS BAUWENS — Specialist birds from this brilliant National flier, "The man with the small basket". Stamina for the North Road, or over the water — winning pure and crossed.

CHURCH VIEW LOFTS

Mr & Mrs D C D Serpell

24 Malts Lane, Hockwold, Thetford, Norfolk IP26 4LA

Phone & Fax: 0842 827873

TRANQUILLITY LOFTS
QUALITY BREEDS QUALITY

Jos Soontjens, Tony Mardon and Staf van Reet.

THE HOME OF THE VERY BEST

Staf van Reet   **Jos Soontjens**
Mol, Belgium   **Wollemgem, Belgium**

Once again a truly exceptional year behind us both for ourselves and so many fanciers who race our bloodlines. We are fortunate to have had full order books for the past two years and also into March 1993 with a few of our stock pairs booked up for the whole of 1993 but then it is widely known we only stock the very best of the original Staf van Reets and the hard to obtain Jos Soontjens. These two strains would take any good fancier to the top in no time at all as we have proven this season.

We have been both successful and trusted racers and breeders for long enough for you to know of our reputation for quality stock but the 1st Fed and 1st Combine winners from our stock has been the best ever from the very top fanciers in the sport to novices and also from fanciers who have been in the sport for some time without success until that is they raced our Staf van Reet and Jos Soontjens pigeons. This season to date (26 August 1992) we ourselves won nine 1sts from ten old bird races, 24 positions from a possible 40 in our section and this after a whole catalogue of disasters during the old bird season.

Two weeks before the old bird season an escaped captive peregrine hawk attacked our Widowhood team killing one and badly injuring two others with five others disappearing never to be seen again and we still won the first race. Result 1st & 2nd Mr & Mrs W A Mardon, vel 1484.48 & 1449.69, 62 miles, "what a pigeon!". Then more disaster as the following Tuesday without our knowledge the field to the rear of our loft was drilled with dressed peas, our Widowhoods were on them in seconds. The result within minutes saw every cock staggering about in a very sad state. Drastic action and they survived the night. A friend who witnessed their behaviour thought they would be dead by morning but they recovered so quickly that we raced them the same Saturday and that

Advertisement continued on next page

was the only race they got beaten for the rest of the season in inland racing.
To put the finish into the disaster catalogue our corn merchant let us down during the last two weeks of the old bird season being unable to deliver our usual mixtures so the feed was changed three times during that time and they still won. Why are we telling you this? Well does this prove to you the sheer quality and courage of our birds? We think most pigeons would have been finished for the year after any one of these events but still our two strains won and won again.

Sheer class and we are proud of every one of them. They did it for us and they will do it for you.

Three young bird races gone and it's business as usual. First race 62 miles, 1st, 2nd, 3rd & 4th Club; second race 62 miles, 1st, 2nd & 4th Club; third race 76 miles, 2nd, 3rd & 4th Club and this with SW or west winds that do not favour us.

STAF VAN REET
"The pigeons with class, speed and courage"

We first went to Staf van Reet in 1987 buying a full round of young birds from his stock loft. We went again in 1988 and 1989 and so you will see all this stock came to us before he sold out his entire loft of breeding stock. It was this stock that had taken Staf van Reet into the limelight drawing upon himself the attention of fanciers the world over. This was the quality pigeon that made Staf van Reet into a household name. It was tragedy when the stock was dispersed out of his hands as we are sure they would have been world famous for many years to come because Staf could have replaced old with young and kept the multi-winners back for the stock loft to add to or replace the older birds. Sadly this can no longer be, however, all our stock came from these pigeons except for a small number which we had from Staf in 1991.

Each one was hand-picked with few crosses introduced. We wanted only the old bloodlines to keep our select family of Staf van Reet pigeons as pure as possible. The new introductions introduced by Staf such as Stichelbaut, Raverstein, Schellens, Raeymaekers and the Soontjens are yet to be tested. This being the "new era" as we have seen it advertised. We do not wish to proceed along these lines and except for one or two directs in 1991 which contain the 1st National Limoges winner of Ravenstein blood we will continue with the old tried and tested outstanding sprinters that came from the years of 1983 to 1989.

These are the highly recommended to you sprint bloodlines that earned Staf van Reet the reputation he now has. Trust us when we say there are no finer quality true Staf van Reets available anywhere in the world today than those we have here at Tranquillity Lofts. We race and breed because we love our pigeons and we are in the sport because it is a sport first and business second place.

Advertisement continued on next page

We will once again retain our price per youngster from direct stock at £40 plus carriage. We are fully aware of the many advertisers who now advertise Staf van Reets but many of them bought their original stock from us and some also ask more for the grandchildren than they paid us for their parents which sounds to us that you get more than a fair deal from Tranquillity Lofts.

Young birds from direct stock (old family) £40 each plus carriage.

JOS SOONTJENS
Sprint Champion 1974-1990

We were introduced to Jos Soontjens by Staf van Reet in 1990 when we were so fortunate to bring back to our stud 40 of these fantastic sprinters. The entire stock of Jos Soontjens has since been sold to Taiwan and are no longer available from Jos Soontjens. The unsuccessful bidder for this incredible family was given free of charge the youngsters in the nest at this time; a round of late breds as consolation.

It was always intended by Jos Soontjens to let us have his best from his best so that we could make his name in Britain and that would enable him to offer his birds into this country so you will understand why we do have his best. Jos Soontjens himself concentrates on racing up to 200 miles but we feel certain that 350 miles is within their capabilities as they have won at 290 miles into the north east as young birds.

The family of Jos Soontjens are very early maturing pigeons that are able to win at highest level as young birds as well as old birds and to any fancier wanting quick results then these are the pigeons for you. A very inbred family of the very highest quality they have won into lofts of some of the best and most well known fanciers in the British Isles and even they were taken by surprise that a new family could come into their lofts and beat out of sight their own successful pigeons. We bought 20 direct from Jos Soontjens and 20 from Staf van Reet bred from his 12 direct Jos Soontjens, we did find in the first season of breeding however that if we paired best to best of the direct Jos Soontjens that some youngsters tended to be very small. A better size came from Jos Soontjens direct to Staf van Reet bred Jos Soontjens and this we recommend to you.

For those of you who have our Jos Soontjens, the pigeons direct from Jos Soontjens are: Saturn, Scamp, Saboteur, Snaffle, Scholar, Sabre, Statesman, Stud, Scout, Select, Scatter, Shadow, Shade, Savoy, Sprocket, Scorpio, Shingle, Scallywag, Shimmer and Saxon. The stock from **Staf van Reet bred pure Jos Soontjens are as follows:** Salute, Style, Supreme, Sizzle, Sagacious, Swallow, Sampson, Silk, Skittle, Season, Scrabble, Scissors, Shilling, Secret, Scarab, Shrimp, Sting, Sherry, Sailor and Signal.

We would suggest to you that a better size of pigeon would be forthcoming if the two lofts of Jos Soontjens could be intermingled where possible.

We personally do not now feel that they are the ideal cross for the Van Reet family of Van Reets. Such a cross may throw up occasional winners, but we feel a good cross should make a greater impact. So we say no to crossing the Jos Soontjens with the Van Reet. We feel a better cross would be Jos Soontjens to another Janssen-based family, the Van Loon, Verheyen for example, or even the original Verheye, which are yet another early maturing family. You may think it unusual for a stud to be so open in making such statements but we do try very hard to put quality pigeons into your hands and will do our utmost to breed you winning stock.

No one will try harder, and we would find it hard to accept that on a percentage basis of stock held that anyone could breed more winners or breeders of winners than ourselves, that's what an obsession to own top quality pigeons does. The pigeon first, money a very poor second.

Advertisement continued on next page

Continued from previous page

Once again the price of youngsters from our Jos Soontjens will be £40 per carriage.

We want to relate a true story to you that happened when a fancier, who will remain nameless, wanted to purchase four of our Jos Soontjens' youngsters. This fancier rang to say he wanted four young birds from our best stock with immediate despatch. He was informed there was a waiting list. He then offered £100 each youngster if he could have them right away. Again he was told there was a waiting list and the price was £40 not £100. The fancier then became very angry informing us that he did not take no for an answer as he was a very successful businessman. He then told me his cheque would be in the post that night for £1,600, yes £400 per young bird! We again said "No" the price was £40 and he would wait his turn. This is a true story. It goes to show how highly fanciers value our pigeons.

Please allow us to show you the quality of pigeon we produce by placing your order as soon as possible. A 20% deposit is required with all orders.

Mr & Mrs W A MARDON

Blithefield, Main Road, Yarburgh, Nr Louth, Lincolnshire LN11 0PW.
Phone: 0507 363304

WARNING

It has come to our attention that false Staf van Reet pigeons are on sale in this country, please check your pedigress. Genuine Staf van Reet pigeons will have a ring number beginning with a first figure of '6'. We have seen pedigrees in which the first figure is '5'. These pigeons have NOT come from the loft of Staf van Reet and were certainly bred outside Staf van Reet's province. Action has been taken against the breeders concerned.—**W A Mardon**

STOP PRESS

Congratulations to Leslie Mairs on his 1st Talbenny YB Derby on 29 August 1992, winning by a clear margin of 20 ypm over 213 miles with a Staf van Reet youngster from our stock. Well done Leslie. Also further congratulations on his 3rd Open Combine, 23,054 birds, 1,556 members with one of our Jos Soontjens' stock.

FRANK CHEETHAM
& SON

SUPER 700-MILERS

Thirty-five birds timed, 59 prizes, in post-war Grand Nationals from: Pau, NFC, 709 miles; San Sebastian, 708 miles; Bordeaux, 608 miles; flying 200 miles further than first drop lofts.

Some outstanding individual performances include: 1947, OLD SCORCHER, 5th Sect E, 28th Open Bordeaux; 1950, 10th Sect E, 67th Open Pau; 1951, 4th Sect E, 7th Open; SUNNYBANK SUPREME 1960, 2nd Sect E, 12th Open, Braithwaite Trophy, RNHU Meritorious Award; 1966, SUNNYBANK FANTASTIC 2nd Sect E, 158th Open Pau, Braithwaite Trophy, RNHU Meritorious Award; 1967, 1st Sect E, 24th Open Pau, Buckley Memorial Trophy, Braithwaite Trophy, Section Cup, RNHU Meritorious Award, Lt-Col Osman Memorial Trophy, for outstanding performance South Road racing. SUNNYBANK INCREDIBLE, 2nd Sect E, 69th Open Pau, Meritorious Award; 1968, DIFFERENT CLASS, 1st Sect E, 168th Open Pau, in worst NFC race ever, only 326 birds timed in British Isles in race time; The Buckley Memorial Trophy (first member to have name on this trophy more than once), Braithwaite Trophy for first bird into Yorkshire for 3rd year in succession (six times in all). 1968, YORKSHIRE BELLE, 3rd Sect E, 302nd Open.

Other outstanding section and open position winners for me are Black Diamond, four times; Rusty; Yorkshire Only One; Arkle; Blue Pearl; Nonchalance, three times; Lady of York; After Dark; Poitiers Hen, twice each; 1953; San Sebastian Queen, 135th Open San Sebastian, scored section and open Pau 1954, plus other prize-winners.

Many of today's top fanciers throughout the British Isles have paid tribute to this family and their progeny, for the high esteem in which they are held today in long distance racing successes. To mention just a few household names are: Geoff Hunt, Westmarsh, Canterbury; T Goddard, Tilehurst; Fisher & Renwick, Up North Combine; Don McCartney, Carnforth, has on more than one occasion timed the farthest flying bird on the winning day from Pau, almost 800 miles; Mr & Mrs Frost, Flockton, Wakefield, did purchase this family in 1969 with the aim of flying Pau NFC, 712 miles, commencing 1972, they have never failed to time since. Their Flockton Flyer has scored in the last six years to 1991. Mr & Mrs Cannon, Godalming, whose successes at Pau are the talking-point on everyones lips; Joe Bradford with his Jubilee Lady, 143rd Open, 21st Open, 1st Open Palamos; J Anderson, Cullybackey, N Ireland, 1st Open YB Irish Derby, Skibbereen, 7,154 birds, with Beechmount Surprise, 1st Open Irish National FC Beauvais, 518 miles, only bird into the whole of Ireland on the day, timed 10.57pm in complete darkness, over 17 hours on wing, only seven birds in three days. Jimmy phoned me to ask if he could name his hen Beechmount Fantastic in honour of my hen SUNNYBANK FANTASTIC.

·This family have won me the Continental Averages on numerous occasions, also Combined Averages all races when due to the severity of the Continental races flown I have finished alone.

I have competed three times in the BICC, one entry each time from Perpignan, 778 miles, taking 32nd, 20th and 7th.

1973, one young bird sent to YB NFC Avranches, 342 miles, timed to take 1st Sect, 248th Open in terrible conditions. I was visited on the third day after timing by Mr Johnston, of Smith Bros & Johnston, Larne, in an attempt to purchase this youngster. Although he was unsuccessful, he asked me to do him a favour and name the pigeon Lionheart, as he had personally experienced the terrible conditions this young cock had to face into to reach home, I did oblige him.

Vandenbroucke-De Weerdt

These are the late Reg Christopher, Fontwell Magna, outstanding family of pigeons, down through his Gold Star. This family have really made a name for themselves locally and by people who have purchased youngsters from me throughout the British Isles. These have turned out to be one of the outstanding

Advertisement continued on next page

Continued from previous page

dual-purpose pigeons that I have ever experienced, winning sprint races, also from the extreme distances, 70-788 miles. *They have won hundreds of prizes for me since I obtained them in 1970.* They have won 1st and 3rd section twice in YB NFC Avranches, also 2nd and 3rd section twice in YB NFC from Avranches, also 2nd and 4th section. 1977, I was 1st Sect, 24th Open YB NFC from Avranches, 342 miles, headwinds all the way, getting stronger the further one went, winning Section Cup, TRPA Trophy for outstanding performance 1977.

Hermans

These are Ward Bros, South Elmsall-Jack Hugill, South Elmsall-based Hermans that I obtained as back-up in sprint races to my Vandenbrouckes, after putting some to extreme distances. They are of the best, their potential has now started to come through for me in the sprint races. I have now added to these by purchasing the best down through Johnny Burton, Knottingley, Hermans, with Hanson Bros Seagull in their breeding, also Anvil Lightning, Anvil Jack, Anvil Fire-eyes.

Once again 1992 has seen another outstanding season beyond my wildest dreams, twenty-five 1st prizes, plus over 155 other positions, 57 of these in Fed, Combine, Amal and Classic races.

My team have held their own in the highest of competition.

A few of my best positions won in central marking competition and Classic club racing 1992 are as follows: I sent nine birds to Angouleme OB Northern Classic, 553 miles, timing two birds on the day, taking 6th Sect B, 6th Open; 21st Sect B, 23rd Open, also timing birds to take good section and Open positions in the top 70. From Le Mans in our two Feds' 400-mile events I sent 13 birds in one Fed, 12 in the other Fed, mainly yearlings. I timed 16 birds within 38 minutes to set a near record for our area for this distance, scoring in Fed, Club, Combine, Amal.

My young bird team once again excelled, scoring weekly, up to the central marking races where it matters. In the only YB race that we had a holding wind, in the Dearne & Dist Fed my birds took 1st, 2nd, 4th, 5th, 7th, 8th Club, 4th, 5th, 7th, 8th, 11th, 14th Fed. The same day my youngsters again came so fast I had 12 in the clock in a short time in the Hemsworth & Dist Fed, taking 2nd, 12th, 13th, 15th Fed, but the icing on the cake came in the Northern Classic YB race from Picauville, 293 miles. I timed three of my eight entries within five minutes to take 1st, 3rd, 5th Sect B, 1st, 3rd, 5th Open, against the top class lofts in northern England, I timed a further two birds to take 34th, 77th Sect B, 36th, 79th Open, winning in the region of £600 plus, also getting the only two birds on the day in Ackworth 2B CC. These were excellent results on a day when returns were mixed with many empty perches and fanciers looking for reasons why??

In my 1992 advert in Squills I advertised some older birds for sale. A fancier from Portsmouth asked me to send down two cock birds. The two year old I sent was not to his liking, the five year old was, but realised he could not break a pigeon of his quality, so both were returned. Much to my delight, as the two year old bred me two hens 1992, one has two 1sts, plus other cards, the other winning a first plus other top cards. The other bird I sent him — Euro Express GB87F52367 — was put back into racing, with the result he came to take 6th Open Angouleme on the day at 553 miles in the Northern Classic.

Due to only losing odd young bird of this excellent young bird team, and having 100% from old bird racing, I will have a few older birds for sale.

I will keep all first round youngsters for my own use 1993. The second and third rounds will all be for sale. If you want to get to the top at any distance, where else can you look to found a loft to get to the top, to compete and keep up amongst the best? Why not try them? Even as a cross? You will get them out of the same pairings that I keep for my own use 1993 racing.

See Squills Year Book 1992, pages 338 & 339, for further details.

1993 young birds for sale from £30 upwards.

Please phone 0977 643045, ask for Frank.
— ALL ENQUIRIES WELCOME —

SUNNYBANK LOFTS
Hillcrest, Waggon Lane, Upton, Pontefract, West Yorkshire WF9 1JT

Advertisement continued on next page

Advertisement continued on next page

of them to take 1st, 3rd, 5th, 6th & 7th. B JOHNSON, Doncaster — flying my pigeons, highest prize-winner in his club. J SWATTEN, Askern, Doncaster — purchasing birds from me 1991, is this year's runner-up Champion YBs. G BUTLER, Stroud — 1st, 2nd Club Worcester; 1st, 2nd, 3rd, 4th Club Worcester, 1st, 3rd, 4th, 5th Combine, 4,000 birds; 1st, 2nd, 3rd, 4th Club Walsall; 5th, 9th, 10th Combine, 4,000 birds; 1st Club Findern. DENNIS WALKER (ace flier), Spennymoor, Co Durham — phoned to say Vapour Trail and Loads of Doe have won between them over £3,000 in the past two years. Other fanciers that have phoned to say how pleased they have been with their birds' performances and have won several 1sts, and includes TONY COLLINGWOOD, Exeter; K LYNN, Dumfries, Scotland: A BEACH and M CANNON, London; N RIVETT, Norfolk; IAN WHEELDON, Stroud; B JOHNSON, Doncaster, R GILMAN, Bradford; and many more too numerous to mention.

REX DOE'S 1992 RESULTS

Club	Fed	Combine
Racepoint — PLYMOUTH OB		
1st, 2nd, 3rd	2nd, 3rd, 9th, 13th, 18th	—
Racepoint — EXETER OB		
1st, 2nd, 3rd	1st, 2nd, 3rd, 4th, 8th 9th, 10th	3rd, 8th
Racepoint — HONITON OB		
1st, 2nd, 3rd	1st, 2nd, 3rd, 9th	14th & 20th
Racepoint — WINCANTON OB		
1st, 2nd, 3rd	1st, 2nd, 3rd, 6th, 9th	6th, 8th, 15th
Racepoint — BASINGSTOKE OB		
1st, 2nd, 3rd	1st, 2nd, 3rd, 4th, 5th, 6th, 8th, 10th, 11th, 14th	4th, 5th, 6th, 11th, 13th, 16th, 18th, 20th, 26th, 27th
Racepoint — SALISBURY OB		
1st, 2nd, 3rd	1st, 2nd, 3rd, 5th, 6th, 7th, 8th, 15th	1st, 2nd, 3rd, 6th, 7th, 8th, 10th
Racepoint — WINCANTON OB		
1st, 2nd, 3rd	2nd, 7th, 10th, 11th, 13th	4th, 8th, 23rd
Racepoint — ASHFORD OB		
1st, 2nd, 3rd	8th, 9th, 12th, 13th, 15th, 16th, 18th, 20th	11th, 12th, 17th, 21st, 22rd
Racepoint — LOWESTOFT OB		
1st, 2nd, 3rd	1st, 2nd, 5th, 10th, 11th, 16th, 18th	20th & 28th
Racepoint — ANTWERP OB		
1st, 2nd, 3rd	3rd, 4th, 9th, 22nd	4th, 5th, 19th
Racepoint — WINCANTON OB		
1st, 2nd, 4th	3rd, 7th & 8th	28th
Racepoint — ANTWERP OB		
1st, 2nd, 3rd	6th, 7th, 8th, 20th & 26th	18th
Racepoint — HONITON YB		
1st, 2nd, 3rd	4th, 7th, 11th, 16th	—
Racepoint — EXETER YB		
1st, 2nd, 3rd	1st, 2nd, 3rd, 7th, 10th, 11th, 13th, 16th	not known
Racepoint — HONITON YB		
1st, 2nd, 3rd	1st, 3rd, 15th	—
Racepoint — HONITON YB		
1st, 2nd, 3rd	6th, 8th	—
Racepoint — WINCANTON YB		
1st, 2nd, 3rd	13th	—
Racepoint — SALISBURY YB		
1st, 2nd, 3rd	6th, 15th, 16th, 17th	—
Racepoint — EXETER YB		
1st, 2nd, 3rd	12th	—
Racepoint — HONITON YB		
1st, 2nd, 3rd	4th, 5th, 6th, 7th, 8th, 9th, 12th, 13th, 16th, 17th	—
Racepoint — BASINGSTOKE YB		
1st, 2nd, 3rd	1st, 2nd, 7th	
One race on the South Road — **HONITON**		
1st, 2nd, 3rd, 4th, 5th, 6th		
SOCIAL CIRCLE — 1st		

Advertisement continued on next page

ALLAN NICHOLS
of DENHOLME

FAUCONNIER ● BUSSCHAERT
JANSSEN ● KRAUTH

Winner of ten 1sts including 1st Pennine Valleys Fed Mangotsfield; 1st Halifax Open race Stafford.

Winner of seven 1sts including 1st Pennine Valleys Fed Cheltenham OB; 1st Pennine Valleys Fed Sartilly (2) OB Championship race.

Winner of thirteen 1sts including 1st Pennine Valleys Fed Nantes; 1st Pennine Valleys Fed Rennes; 2nd Pennine Valleys Fed Nantes 1990; 21st Chris Catterall Championship Nantes, 14,531 birds.

1989 results — Thirteen 1sts and eleven 2nds and Combined Av, Combined Inland Av, YB Av, YB Knock-Out, OB Knock-Out, Points Cup winner, highest prize-winner for 8th year out of 10. First Luddenden Foot HS Picauville 500 Nominated Yearling race, winning £500 prize money + £44 pools; 1st, 2nd, 3rd & 4th Slaithwaite Open race Stafford, 36 members sent 270 birds, 50 ypm in front, winning £69; 1st, 9th, 10th & 15th Halifax Open Stafford, 78 members sent 629 birds, winning £82; 2nd Queensbury Open Dorchester, 54 members sent 203 birds, winning £67; 21st Chris Catterall Championship Race Nantes, 14,531 birds. **In Pennine Valleys Fed** — First & 17th Nantes OB Championship, 455 birds, £76 pools plus trophies for best 2-Bird Av, lowest winning vel of all Channel races; 1st & 8th Mangotsfield OB, 2,215 birds; 3rd & 4th Cheltenham OB, 1,489 birds; 3rd, 19th, 27th & 30th Cheltenham YB, 2,081 birds; 3rd & 12th Stafford (2) YB, 2,653 birds; 7th, 22nd & 28th Cheltenham OB, 1,713 birds; 8th, 19th & 25th Rennes OB Championship, 717 birds, £29 pools; 9th & 13th Worcester (2) YB, 2,619 birds; Runner-up Channel Av, 3rd Chris Catterall Trophy Points. **In Ogden HS** — Eight 1sts, five 2nds, five 3rds, eight 4ths; Combined Av, Combined Av all races, Championship Av, Shield for Champion OB for the second year running, highest prize-winner again. Channel Av for second year running, OB Inland Av for 14th year out of 16. **In Haworth FC** — Twelve 1sts, ten 2nds, ten 3rds, four 4ths; Channel Av for second year running, OB Inland Av for 9th year out of 11.

1990 results — 3rd, 4th & 13th Halifax Open race from Stafford, 72 members sent 580 birds, winning £61 pools + prize money; 5th, 10th, 13th & 15th Queensbury & Dist Open race from Dorchester, winning £175 pools + prize money. **In Pennine Valleys Fed:** 1st & 13th Cheltenham (2) OB race, 1,478 birds; 2nd & 6th Nantes OB Championship race, £130 pools, beaten by less than 1 ypm, winning the Tetley Trophy for best 2-bird average for second year running; 4th, 5th, 19th & 20th Stafford OB, 1,936 birds; 4th, 8th & 22nd James Oven Memorial race Mangotsfield, 409 birds, winning £20; 4th, 5th & 15th Dorchester YB Championship race, 1,337 birds; 6th & 7th Sartilly (2) OB Championship race, £35 pools; 9th, 11th, 15th, 19th & 21st Worcester YB, 1,564 birds; 10th, 16th & 24th Cheltenham (3) OB race, 1,258+ birds. 12th & 13th Sartilly (1) OB Championship race, 860 birds; 13th, 22nd & 25th Dorchester OB Championship race, 1,740 birds; 13th Worcester (1) YB race, 2,180 birds; 14th & 20th

Mangotsfield (3) OB race, 1,369 birds; winner of the James Oven OB & YB Av; 3rd Championship Av; 3rd Channel Av; 3rd Chris Catterall Trophy Channel Points. **In Ogden HS:** Nine 1sts, nine 2nds, eight 3rds, seven 4ths; Combined OB Av for third year running; winner of YB Cup for 2nd year running; OB Inland Av for 15 years out of 17; Channel Av for third consecutive year; Championship Av for third consecutive year; highest prize-winner for 15 years out of 17. **In Haworth FC:** Eight 1sts, twelve 2nds, five 3rds, six 4ths; Channel Av and Championship Av for third consecutive year; Combined OB & YB Points Shield for fourth year running; OB Inland Av for tenth year out ot 12; highest prize-winner for 9th year out of 11.

1991 results — Twenty-seven 1sts, twenty-four 2nds, twenty-one 3rds, nineteen 4ths. **In Pennine Valleys Fed:** 1st, 13th, 15th & 21st Sartilly (2) OB Championship race, 678 birds, winning £89 pools; 1st, 8th & 21st Cheltenham (1) OB, 2,630 birds; 2nd, 12th & 30th Sartilly (1) OB Championship race, 925 birds, winning £66 pools; 2nd, 6th, 7th, 8th & 10th Mangotsfield (3) OB, 1,213+ birds; 2nd & 6th Stafford (2) YB, 2,800 birds; 6th, 19th & 25th Cheltenham (3) OB race, 1,508 birds; 7th, 20th and 23rd Nantes OB Championship race, £33 pools; 7th & 28th Worcester (1) YB race, 2,413 birds; 11th & 13th Mangotsfield YB, 1,674 birds; 11th Mangotsfield (2) OB race, 1,718+ birds; 11th James Oven Memorial race, Newton Abbot (YB), 690 birds, winning £69 pools + prize money; 13th & 24th Cheltenham (2) OB race, 1,491+ birds; 13th Worcester (2) YB race, 2,356 birds; 13th, 16th & 23rd Rennes OB Championship race, 637 birds; 16th, 26th & 27th Stafford OB race, 1,958 birds; runner-up Championship Av and Channel Av; runner-up Chris Catterall Trophy Channel Points; 2nd Queensbury & District Open Race Guernsey, beaten by decimal; 3rd & 5th Ambler Thorn Open Training Race Stafford (3), 67 members sent 602 birds; 4th Ambler Thorn Open Training Race Stafford (2), 49 members sent 400 birds; 4th & 7th Ambler Thorn Open Training Race Stafford (1), 54 members sent 427 birds, winning a trophy for the best average over all three races, plus £106 pools and prize money. **In Haworth FC:** Combined OB Av & Channel Av for fourth year running; Combined OB & YB Points Shield for fifth year running; OB Inland Av for 11th year out of 13; Highest prizewinner for the tenth year of out 12. **In Ogden HS:** Combined OB Av & Channel Av for the fourth consecutive year; OB Inland Av for the 16th year out of 18; Highest prizewinner for the 16th year out of 18.

1992 results — *at time of going to press, with three races yet to fly:* Twenty-five 1sts, twenty-two 2nds, twenty-three 3rds, twenty 4ths. **In Pennine Valleys Fed:** 1st, 9th & 14th Fed Stafford OB, 2,032+ birds; 1st, 8th, 14th, 18th, 23rd & 27th Fed Mangotsfield OB, 2,331 birds; 1st, 14th & 16th Fed Mangotsfield OB, 1,747 birds; 2nd, 13th, 20th & 27th Fed Cheltenham OB, 1,378 birds; 4th, 6th, 17th & 30th Fed Sartilly (2) OB Championship race, 764 birds, £32 pools; 4th, 7th, 10th, 11th & 16th Fed Mangotsfield (3) OB, 991+ birds; 5th Fed Satilly (1) OB Championship race, 746 birds, £90 pools; 5th, 23rd, 27th Fed Worcester (2) YB, 2,487+ birds; 6th, 15th, 26th Fed Worcester OB, 1,937+ birds; 7th, 12th & 17th Fed Cheltenham (1) YB, 2,117+ birds; 10th Fed Rennes OB Championship race, 714 birds; 12th, 17th, 19th & 20th Fed Cheltenham (3) OB, 1,465 birds; 13th & 26th Fed Dorchester OB Championship race, 1,730 birds; OB Inland Average. *Also: 2nd, 5th & 6th Halifax Open Race, Stafford, 114 members sent 1,047 birds, winning £61; 3rd, 6th & 19th Keighley & District Open Race Cheltenham, 56 members sent 330 birds, winning £30; 4th, 7th, 9th & 14th Ambler Thorn Open Training Race (3) Stafford, 56 members sent 447 birds, winning £10 prizemoney; OB Inland Average Haworth FC for the 12th year out of 14; OB Inland Av Ogden HS for the 17th year out of 19.*

1993 youngsters from stock pairs or race birds, £25-£50 each *plus carriage.*

Full pedigrees supplied. Also a few OBs available.

ALLAN NICHOLS
6 New Street, Denholme, Bradford, West Yorkshire BD13 4AE
Phone: Bradford (0274) 835152

The Rock Lofts

Home of long distance winners this last 40 years.

Seven times winners of 1st Sect in National Flying Club races.
Old birds — four times 1st Sect Pau, 683 miles, three times 1st Sect Nantes, a record at this distance, racing against the best of long distance lofts.
Winners of four RPRA Centre Awards at over 500 miles, and responsible for five more for other fanciers.

| 1st Sect, 46th Open Pau, vel 666, 4,585 birds. | 1st Sect, 7th Open Pau, 6,928 birds. | 2nd and 3rd Sect Pau; 106th Open Pau. |

Our wins from NFC Pau, 683 miles, average of four birds per race: 1st Sect, 5th Open Pau; 5th, 8th, 9th Sect, 199th, 231st and 259th Open Pau; 9th, 10th and 27th Sect Pau, 7th Sect; 282nd Open Pau, 24th Sect Pau; 7th Sect, 284th Open Pau; 9th Sect, 259th Open Pau; 1st and 5th Sect, 15th, 133rd Open Pau; 3rd Sect, 364th Open Pau; 27th Sect Pau; 8th Sect, 231st Open Pau; 17th and 18th Sect Pau; 6th Sect, 327th Open Pau; 1st Sect, 7th Open Pau; 2nd Sect, 106th Open Pau; 3rd Sect, 323rd Open Pau; 1st Sect, 46th Open Pau; 9th Sect, 270th Open Pau; 11th and 15th Sect, 151st and 205th Open Pau; with up to and above 6,000 birds some races.

Also they win the long distance races for others: four 12in x 8in sheets of wins for other fanciers can be had on request.

YBs available from stock and race birds, 500 miles to 683 miles, from £30 each.

See Squills 1987, 1990 and 1991 for other stock details.

Visitors welcome any time, phone first.

E FOX & SON
"Rocklea", Buxton Road
Bakewell, Derbyshire DE4 1DH
Phone: 0629 814305

Advertisement continued on next page

The Rock Lofts

Winners at Saintes, 518 miles; Angouleme, 529 miles; Bordeaux, 581 miles; Bergerac, 586 miles; Pau, 683 miles. Winners on fast or slow races for ourselves and others.

The six birds below are typical of our family, **a combination of S P Griffiths, Delmotte, Hansenne, Osman, and a few of A H Bennett used as a cross, now breeding pigeons to win good positions in the long races.**

1st Sect, 15th Open Pau, 4,773 birds.

1st Sect, 25th Open Nantes, 9,685 birds.

1st Fed Saintes, breeding winners.

2nd Sect, 170th Open Bordeaux NFC, 3,776 birds; 5th Sect, 19th Open Bergerac Mid Nat.

3rd Sect, 237th Open Bordeaux NFC; 2nd Sect Nantes NFC; 8th Sect, 62nd Open Angouleme.

Sire and grandsire of winners to Pau.

The above six birds all related and breeding winners.

E FOX & SON
"Rocklea", Buxton Road, Bakewell
Derbyshire DE4 1DH
Phone: 0629 814305

WELSH SUPREME LOFTS OF
KEN
DARLINGTON
GILBERT LANE NURSERIES
Barry, South Glamorgan CF6 3BD
Phone: 0446 738733
JANSSEN x VANDENBOSCH
LEADERS OF THE NEW GENERATION
TOP FLIER WELSH NORTH ROAD FED
1,500 members, 1989, 1990 and 1991. — Biggest Fed in UK.
Also top flier Cardiff Fed 1989, 1990 and 1991.
Top flier Centre East South Fed 1986, 1987, 1988, 1989,
1990, 1991 and 1992
OVER FOUR HUNDRED 1sts IN SEVEN YEARS

Ken Darlington holding the Silver Pigeon Trophy, only the second flier to win the Silver Pigeon three years in succession. To win this trophy only top three positions in the Fed count (1,500 members).

JANSSEN x VANDENBOSCH
KEN DARLINGTON

1st WELSH GRAND NATIONAL, 1st COMBINE (9,488 birds) — QUEEN'S CUP WINNER 1986

TOP SPRINT CHAMPION 1990 ALL WALES — W G BOWEN CUP

TOP YB CHAMPION 1989 ALL WALES

No other loft in Wales has ever won the three top cups in Welsh Homing Pigeon Union.

TO WIN THESE CUPS A PIGEON MUST BEAT HALF A MILLION PIGEONS

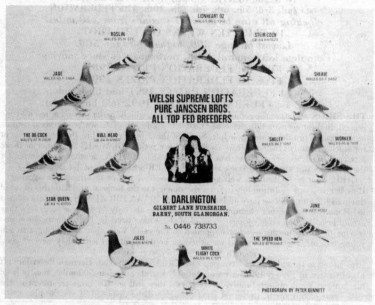

SPENCE & NICHOLLS
"THE WELSH WIZARDS"

Winners of twenty-five 1st Federation from 1982 to 1987 in both the Cardiff Federation, and Welsh North Road Federation with up to 15,000 birds competing. *Records set at all levels.*
1st FED ALL WALES TWO WEEKS RUNNING
1st & 2nd CENTRE COMBINE BY OVER 30 ypm
1st, 2nd, 3rd, 5th, 6th, 7th, 9th, 10th, 11th FEDERATION
clocking all nine birds in 57 seconds from 240 miles.
1st CLUB TEN WEEKS ON THE TROT
Lowest birdage 300, highest birdage 500.
Complete loft removal 1988, performances since moving:
FIVE 1st FEDERATION PRIZES 1989
FIVE 1st FEDERATION PRIZES 1990
SIX 1st FEDERATION PRIZES 1991 (old birds only)
FIVE 1st FEDERATION PRIZES 1992 (old birds only)

THE BIRDS — Before the removal sale of our birds, which by the way set a new all-time record, highest ever auction price per bird ever sold in the UK, we bred several rounds of youngsters to take to our new loft. Some of these were raced but the best were kept for stock. Already of the birds we raced, we have retired BLACKIE II, winner of three 1sts Fed; BOB, winner of three 1sts Fed; THE LAST TROJAN, another Fed winner. Of the birds we put straight to stock we have SINGER'S SON, sire of 1st Fed; SINGER'S GIRL, dam of 1st Fed. As well as all these we have a host of others just as good.

However, it is our ambition to win from the shortest to the longest, and with this in mind during 1988 and 1989 we have added to our birds 70 VAN BRUAENES all closely related to his many long distance champions. However, there are only 34 of these birds left, they are all breeding top class youngsters — 14 of them have bred 1st Fed winners. Now, this is a point worth noting — what happened to the rest? Well, they weren't kept to sell youngsters off, they weren't sold for others to be disappointed with, they weren't good enough for us, so they were killed.

Also added to these birds we have imported 70 JANSSENS from Herman Beverdam off his birds that in 1990 & 1991 have won not only sprints but up to 500 miles, mostly from his Janssen Vandenbosch family. Don't worry, they will get the same treatment as the rest of our birds, those not good enough will be disposed of via the bin.

— THE ELECTRICK HEN 84 —   — SURE BET —

A grand-daughter of Van Bruaene's Electrick of 67 when she was paired to one of our grandsons of Barcelona II. She bred First Choice, a winner of four 1sts plus 1st Fed, and Bird of the Year as a yearling, plus 1st Fed again as a two year old, etc, for Spiller, Iles & Billen. Nestmate of The Electrick Hen bred our 1st Fed Crieff winner 1991.

Typical of the recent imports from Herman Beverdam. Janssen Vandenbosch.

Advertisement continued on next page

Continued from previous page

— THE BIB —

A brilliant producer, we can trace seven generations of 1st Fed winners bred down from this cock. Birds like The Singer, Welsh Queen, Blackie II, Young Bib, all big winners with up to 12,000 birds competing. The Bib is a son of Green Eye, the blue Janssen Bros cock when paired to La Bonne, the pied hen of the old Bollaerts (Vandenbosch) family that won 37 prizes racing Natural.

— HERMAN —

Bred by H Beverdam, bought by us as a late bred in 1983. He is sire, grandsire, great-grandsire, etc, of many, many Fed winners. Howells & Bayliss of Nantymoel have a grandson off him that is responsible for 13 Fed winners alone. Herman is one of the best pure Janssen producers in the country, both parents bred by Janssen Bros, Arendonk.

— BLACKIE II —

Winner of three 1sts Fed, club, 2-bird and open wins also, plus colour TV, etc. He is a son of Blackie, twenty-three 1sts, five 1sts Fed, etc. Dam of Blackie II is Sissy, own daughter of Welsh Queen and Herman. Welsh Queen another super daughter of The Bib.

— THE PHARAOH —

A grandson of Van Bruaene's Den Fynen. The Pharaoh bred a Fed winner in his very first nest. We also have at stock his nestmate that has also bred a Fed winner. Typical of our Van Bruaene stud cocks.

After we have kept the 1993 first round of youngsters for ourselves all young birds will be for sale.

They will be £50 each or £200 for six. These will be from our stock birds including Federation winners and breeders of Federation winners, *OR* £100 each and £400 for six bred from the multiple Federation winners and breeders of multiple Federation winners like HERMAN, THE BIB, BLACKIE II, etc.

Apply in confidence to:

SPENCE & NICHOLLS
5 Dorset Avenue, Barry
S Glamorgan CF6 6BH
Phone: 0446 730850 or 736547

Advertisement continued on next page

Advertisement continued on next page

Continued from previous page

SCHLEPPHORST

Janssen-Meuleman. We have a number of direct Schlepphorst birds which contain most of the top winning bloodlines. They are aways in demand, winning through to 400 miles.

☆

KROUWEL-POLLMAN

Jan Aarden, long distance — we have direct birds of this partnership. We have the same bloodlines as winner of 1st Open Dutch National Barcelona, 758 miles, beating 8,163 birds. Our bloodlines include De 91, De 27, Prinses, Oude 49, Prince, etc.

☆

MAURICE DELBAR

We only have a small stud of Delbars but most are multi-prize-winners or breeders of 1st prize-winners. The Delbars are a good all-round family, good for sprint-middle or long distance.

☆

LEFEBRE-DHAENENS

At PJ LOFTS we have a large number of direct Lefebre-Dhaenens plus all the best of the '70s original imports, sprint, middle distance. We have had Fed winners reported from 65 miles to 540 miles with our Lefebre-Dhaenens.

☆

BUSSCHAERT

Only recently introduced, we have birds very close to the originals. We have the bloodlines of Pluto, Klaren, Sooten, Crayonne, Sylvie, Broken Toe, Lady Killer, Coppi Hen, Clare, Schiavon, Devina, 505, Taco, Petra, Champion Little Black, etc.

☆

BEVERDAM JANSSEN direct

Sprint, middle distance. Very good family, winning throughout the country. We have most of the top bloodlines.

☆

STAF VAN REET

Very good sprint family. We only have a small family of Van Reets but they are producing many winners. We have all the top bloodlines of this family including Daniel, Vosse Wittekop, Dikke Prinz, Tienneke, De 190, etc.

Advertisement continued on next page

Continued from previous page

We have many reports of winners every week with our birds, from all round the country.

☆

Young birds from all families start at
£30 each; six for £150; ten for £230.
£50 each; six for £250.
£100 each; six for £500.
A good reduction for large orders.

☆

Ideal Christmas or birthday present —

PJ LOFT GIFT VOUCHERS from £5.

☆

*All birds bought go into a £1,000 prize draw again for 1993,
including prize for the best performances with our young birds.*

☆

Also for sale — "LOFT, STOCK & BARREL"
by B McNicholas. £12.50 plus £1.50 p&p.

For any details please contact:

Paul Newbold
PJ LOFTS

262 Humberstone Lane
Leicester LE4 7UN
Phone: 0533 694630
VENUES FOR AWAY SALES WANTED

Advertisement continued on next page

My local club, Balby HS, is probably the biggest club in the Fed (formerly 66 members strong), sending up to 600 birds but, to be fair, now about 40 members. My birds have broken all club records, being premier prizewinner: 1975-76-77-78-79-80-81-82-84-85-86-87-88-89-90-91. My record during the past seven years is as follows: 1985, twenty-eight 1sts, twenty-four 2nds, twenty 3rds; 1986, twenty-eight 1sts, twenty-one 2nds, sixteen 3rds; 1987, twenty-nine 1sts, twenty-nine 2nds, twenty-eight 3rds; 1988, thirty-two 1sts, twenty-three 2nds, thirty-one 3rds; 1989, forty 1sts thirty-one 2nds, thirty-four 3rds; 1990, forty-one 1sts, twenty-seven 2nds, thirty-two 3rds; 1991, twenty-eight 1sts, twenty-two 2nds, twenty-two 3rds.

Since 1967, the Hermans are without doubt the finest sprint pigeons to come into South Yorkshire and still are to this day. No other strain of pigeon has won more than the Hermans in South & West Yorkshire in sprint races.

I, along with my father, purchased some Karel Hermans in 1967 and also in the early '70s. I have had them ever since and have blended them over the years into a family of sprint pigeons second to none. There is only one way to describe them and that is "absolutely brilliant". They were breaking records in 1975 and are still breaking them today in club, Fed and Open competition. I never cease to be amazed at their fantastic performances, year after year. Do not take my word for it, ask anyone about the Hermans in South & West Yorkshire. I have tried all kinds of top sprint pigeons, but none of them can live up to the Hermans and, believe me, they have to be top class to win in this area. I regard the top fanciers in South & West Yorkshire and the Doncaster area to be as good as any fanciers in the country. They certainly do not get the publicity and recognition which they thoroughly deserve in this hotbed of sprint racing. With such competition you can rest assured that, if your birds win here, then they can win anywhere. I know, because I have them to beat every week.

I have listed just a few of the best results put up by these "record breaking sprinters" against, at times, up to 600 birds in my club and 5,200 in the Fed. Just study them; they may or may not be the best performances ever against top comeptition, but they will certainly take some beating. I will leave that for you to decide and I will let my birds do my talking for me: 1st, 2nd, 3rd, 4th, 6th Hemsworth Open, 500 birds; 1st, 2nd, 3rd, 4th, 5th Hemsworth Open, 433 birds; 1st, 2nd, 3rd, 4th, 5th, 6th, 7tyh, 8th, 10th Club Wallingford, 1st, 4th, 5th, 7th, 8th, 9th, 12th, 16th, 18th Fed; 1st, 2nd, 3rd, 4th, 5th, 6th Conisbrough Open (open to all Yorkshire): 1st, 2nd, 3rd, 4th, 14th, 18th, 21st, open to all Yorkshire, 1,000 birds; 1st, 2nd, 3rd, 4th, 6th, 7th, 9th Club, 426 birds, 1st, 2nd, 3rd, 4th, 12th, 14th, 20th Fed, 4,835 birds; 1st, 2nd, 3rd, 4th, 5th, 7th, 8th, 9th, 10th Club, 1st, 2nd, 6th, 9th, 12th 18th Fed, 3,000 birds; 1st, 2nd, 3rd, 4th, 5th, 6th Club, 578 birds, 2nd, 3rd, 8th, 14th, 15th Fed, 5,710 birds; 1st, 2nd, 4th, 5th, 6th, 7th Club, 550 birds, 1st, 2nd, 4th, 5th, 6th, 10th Fed, 5,400 birds; 1st, 2nd, 3rd, 4th, 5th, 8th, 10th Club, 400 birds, 1st, 2nd, 3rd, 4th, 5th, 8th, 10th Fed, 3,500 birds; 1st, 2nd, 4th, 5th, 7th, 8th, 9th, 10th Club, 379 birds; 1st, 2nd, 6th, 9th, 16th Fed, 3,167 birds; 1st, 2nd, 3rd Club, 1st, 2nd, 3rd Fed, 3,000 birds; 1st, 2nd, 3rd, 4th Club, 510 birds, 1st, 4th, 5th, 8th Fed, 5,000 birds; 1st, 2nd, 3rd, 4th, 5th, 6th Club, 333 birds, 8th, 10th, 12th, 16th, 19th, 20th Fed, 3,500 bird; 1st, 2nd, 3rd, 4th, 5th, 6th, 7th, 9th Club, 610 birds, 7th, 8th, 13th, 19th Fed, 4,000 birds; 1st, 2nd, 3rd, 5th, 7th, 8th, 9th Club, 1st, 7th, 9th, 13th, 16th, 19th Fed, 2,100 birds; 1st, 2nd, 4th, 5th, 7th, 8th, 9th Club, 4th, 8th, 12th, 15th, 18th Fed, 1,825 birds; 1st, 2nd, 4th, 6th Club, 300 birds, 1st, 2nd, 4th, 6th Fed, 5,000 birds; 2nd, 3rd, 4th, 5th, 6th, 7th Club, 500 birds, 3rd, 4th, 5th, 6th, 8th, 9th Fed, 3,900 birds; 2nd, 3rd, 4th, 5th, 7th, 8th Club, 5th 6th, 8th, 9th, 14th, 16th Fed, 3,216 birds; 1st, 2nd, 3rd, 4th, 5th, 6th, 7th, 8th Midweek Club, 272 birds; 1st, 2nd, 3rd, 4th, 5th Midweek Club, 163 birds; 1st, 2nd, 3rd, 4th, 5th, 7th, 8th Midweek Club, 168 birds; 1st, 2nd, 3rd, 4th Midweek Club. This is the Fairway Midweek Club, one of the strongest midweek clubs in the area, with some of the best fanciers around. In this club, with approximately 60 members, I have many times 1st, 2nd, 3rd Club. 1st, 2nd Clowne Open from Banbury.

Price of young birds £35 and £50 each.

Alwyn Paddey, 23 Goldsmith Road, Balby, Doncaster DN4 8LZ. Phone: 0302 858508, after 6pm

The Third Generation — 1992

July 1992 saw a further visit to Mol adding more sons and daughters from Staf's current top stock birds.

Young birds to race 1993 £40 each or £200 for six or £300 for ten.

All with guaranteed direct Staf van Reet parents.

"The nearer the root the sweeter the fruit."

5th Fed Dorchester; three 1sts and 5th Fed Dorchester Champ.

Hession blood, dtr of Hotfoot.

Four 1sts including 1st Fed Cheltenham.

Son of Toey, sire and grandsire of winners.

FINALLY — THE HESSION SVRs

Bert brought these into this country in 1985. Having seen as YBs in 1986 Bert's now legendary star performers, we realised we could be in at the beginning of a new era in sprint racing. With this in mind we purchased in 1987 and 1988 children from all his star performers, ie Toyboy, Playboy, Hotfoot, Mister Magic, Genopte, Kojak, White Eye, etc. Plus children from his star breeders.

1993 YBs from £50 each, £270 for six.

Advertisement continued on next page

Belg 88. Dtr of Elf, six 1sts.

1991 direct Staf van Reet.

Hession blood, son of Toyboy, fifteen 1sts.

Four 1sts and 1st Fed Cheltenham.

WE TEST THEM — YOU BENEFIT!!

Flying in one club — Keighley & District HS with over 30 members — the Van Reets have ensured we have been highest prizewinner 1988 — 1989 — 1990 — 1991 — 1992 (to date).

These performances achieved despite full-time farming.
THESE PIGEONS ARE IMPACT PIGEONS. — All above orders 10% deposit. Contact:

Richard or Sue Grayston
Dawslack Cottage, Dawslack Farm, Low Utley, Keighley, West Yorks BD20 6DL.
Phone: 0535 669038

Visitors by prior arrangement please.
Stop press: *Just purchased a small select stud of pure Jos Soontjens direct from Belgium. 1993 YBs — prices on application.*

Steven van Breemen pigeons

stand for 30 years of highly successful racing.
The concentrated gene pattern of these pigeons is responsible for a powerful passing-on capacity that goes beyond your imagination.

Presenting the superstar of the loft:
H82-448368 THE GOOD YEARLING
★ **1st Semi-National Chateauroux, 650km, 8,500 birds.**
★ **2nd National Ace Yearling 1983**
★ **One of world's finest breeders.**

Steven van Breemen pigeons — mainly based on the Good Yearling family — produce great over nine generations. Imagine . . .

Advertisement continued on next page

Continued from previous page

THE ART OF BREEDING
developed by Prof Alfons Anker
and Steven van Breemen:
THE NEW WAY OF RACING PIGEONS

1992 results of
Steven van Breemen's
GOOD YEARLING FAMILY

Total long distance old bird results:
Regional average 1,200 birds/600km each race (four races total):
1−2−2−3−5−6−7−7−7−10−11−20−22−23−26−32−33−33−43−47−59−50, etc.
1st Champion Long Distance Club.

National average 8,000 birds/600km each race:
1−13−19−30−33−38−41−43−50−79−95−99, etc. — **4th Champion NABvP.**

Special result:
**1st National Chateauroux, 660km, 5,523 birds, six minutes ahead. H91-5055626
direct son of H82-448368 The Good Yearling.**

Provincial young bird results:
100km, 5,799 birds: 3−4−5−6−8−10−72−76−83−84−98−111−121, etc.
135km, 6,760 birds: 8−13−14−24−27−33−37−39−42−45−49−54−55−60−61, etc
225km, 8,052 birds: 31−34−35−39−40−48−49−52−138−139−156−160, etc
270km, 7,341 birds: 7−12−13−29−37−39−41−54−55−56−84−102−109−112, etc
330km, 5,801 birds: 3−6−7−10−21−37−50−53−60−100−101−105−117−121, etc
330km, 6,602 birds: 13−14−15−18−20−22−23−63−67−77−80−92−102, etc
270km, 4,907 birds: 4−5−7−10−11−12−13−15−17−20−23−24−26−27−33−34, etc
540km. 4.330 birds: 2−7−8−16−17−18−20−25−27−31−39−43−51−64−88, etc

**National Orleans 1992 against about 30,000
birds, we place 15 birds under the first 150
positions!!**

*Do you want to receive detailed information
about this loft?*

*Interested in pigeons with guaranteed high
breeding qualities?*

Write or phone:
Steven van Breemen
Leeghwaterstraat 68
1221 BH Hilversum, Holland
Phone & Fax: 010-31-35-830285

ROBERT CARETTE
Leeuwerikstraat 132, 8400 Oostende
Phone: 059/80 38 40

Robert Carette Barcelona — Only two birds sent each year:
1985 — 489th & 1,840th, 8,647 birds. 1986 — 1730th, 8,846 birds. 1987 — 205th & 1,383rd, 8,915 birds. 1988 — 62nd & 1,097th, 10,186 birds — 1st Barcelona Club Nieuwpoort, 12th Golden Wing Award. 1989 — 824th, 11,656 birds. 1990 — 70th & 713th, 12,283 birds.

National Marathon Results: 1987 — 18th. 1988 — 11th. 1989 — 1st Belgian Pigeon Futurity Race in Japan. 1990 — 34th; 13th Champion of Belgian Big Fond; 1st Champion Union Club Gistel; 1st Europa Cup Barcelona.

DE "MARATHON - VLIEGERS"

1986 Narbonne	Nat.	3.992 D. 111
1987 Montauban	Nat.	4.017 D. 162
1988 Marseille	Internat.	5.389 D. 321
1990 Dax	Nat.	3.236 D. 52
	Internat.	4.730 D. 63
Perpignan	Nat.	3.306 D. 104
	Internat.	10.444 D. 193

"PLASTRON"
3237031-83

1985
Angoulème Reg. (N) 1
Barcelona (Duivinnen)
Reg.   4
Prov.   12
Nat.   32
Internat. 83

"MIJO"
3253038-84

VAN

"BARCELONA"
3250656-85

CARETTE ROBERT
LEEUWERIKENSTRAAT 132
8400 OOSTENDE
TEL. (059) 80 38 40

"MONTAUBANTJE"
3250661-85

1987 Barcelona	Nat.	8.915 D. 205
	Internat.	21.545 D. 303
1988 Barcelona	Nat.	10.186 D. 62
	Internat.	21.194 D. 161

1987 Montauban	Nat.	4.017 D. 52
1988 Narbonne	Nat.	6.167 D. 172
1989 Montauban	Nat.	4.685 D. 514
1990 Narbonne	Nat.	4.420 D. 98

"DIKKOP"
3250687-85

1990
Barcelona	Nat.	12.283 D. 70
	Internat.	28.128 D. 184
Perpignan	Nat.	3.306 D. 157
	Internat.	10.444 D. 315

"VERMOTE"
3352859-85

Vader van 1° Belgische Duif
Internationaal Futurity Race
Japan 1989

VANDEVELDES AND BUSSCHAERTS

from the winning lofts of

MANSON BROS

of Hartlepool

The Vandeveldes are a winning strain that we have cultivated for the last 30 years. They have excelled at distance racing for us and other fanciers.

Below are examples of this winning strain.

B C, 19th and 121st UNC Bourges, 14 hours on day.

B H, 4th UNC Bourges, 7,809 birds, 16 hours on day.

B H, 5th and 39th UNC Bourges.

B W H, 2nd UNC Bourges, 7,690 birds, 16 hours 20 minutes on day

M H, 30th and 196th UNC Bourges.

B C, 44th UNC Bourges, 14 hours on day.

Advertisement continued on next page

The Busschaerts are a more recent introduction from some of the top Busschaert lofts in the country. These have been an instant success for us. Below are examples of our winning Busschaerts.

Four 1sts Club, 1st, 2nd, 2nd, 9th, 17th Fed.

Three 1sts Club, 1st, 5th, 6th, 6th, 16th Fed.

Four 1sts Club, 1st, 1st, 6th, 7th, 9th Fed.

Three 1sts Club, 2nd, 2nd, 7th, 8th, 14th Fed.

Six 1sts Club, 2nd, 2nd, 4th, 9th, 12th, 16th Fed.

Three 1sts Club, 4th, 5th Fed.

Four 1sts Club, 1st, 3rd, 3rd, 12th, 12th, 14th Fed.

Three 1sts Club, 3rd, 4th, 7th, 10th, 10th, 10th, 14th Fed.

Two 1sts Club, 4th, 4th, 6th, 8th Fed.

1st, 2nd Fed, 21st, 34th UNC Lillers.

23rd, 163rd UNC Provins, 135th UNC Lille, 80th UNC Melun, 223rd UNC Clermont.

21st UNC Melun, 169th UNC Abbeville, 157th UNC Clermont, 141st UNC Provins.

3,000 + birds go in this Fed.

Advertisement continued on next page

BUSSCHAERTS

1st Fed Peterborough

1st Fed Folkestone

18th UNC Abbeville

16th UNC Abbeville

Flying in **one club only, Hartlepool WCHS** (23 members), our clubs wins in last four years are: thirty-two 1sts, thirty-one 2nds, twenty-three 3rds, twenty-five 4ths. Our club wins 1992, OBs only (16 races): seven 1sts, four 2nds, two 3rds, three 4ths.

Our recent **Up North Combine** positions: 5th, 11th, 18th, 20th, 21st, 21st, 30th, 34th, 39th, 39th, 69th, 72nd, 82nd, 91st, 108th, from 326 to 551 miles. **Money won in Hartlepool WCHS and three Championship Clubs from 1988 to 1991 season is £7,100.**

We have a limited number of young birds for sale, prices from £30 plus carriage. Sorry none sold locally.

Apply to:

MANSON BROS
12 Dorchester Drive, Hart–Station, Hartlepool, Cleveland TS24 9QY
Phone: 0429 233255

The loft of Manson Bros has excelled at distance racing in the UNC. Since introducing the Busschaert they have excelled at all distances. A fair deal is assured.—**UNC**

Advertisement continued on next page

ORBITAL

QUALITY always wins
QUALITY always shows

WE SELL QUALITY

The very best of
Janssens
Aliens

Direct **Hartogs**

We are also selling youngsters from our racing loft.
Over thirty-five 1sts to date this season including 1st SMT Combine Le Mans [5,363 birds].
Phone now, mature youngsters for sale.

Roger, 0734 477149

Advertisement continued on next page

Advertisement continued on next page

1948 — 1992
HENRY HADFIELD & SON

Having always been interested in the long distance races over the years I have blended together pigeons from different top long distance lofts and by pursuing a policy of selective line breeding I now have a strain which I consider to be my own.

Thêy score from all distances but excel from the Channel at 300 miles plus. The harder the day the better they like it.

They are also a good looking family and over the last few years they have won hundreds of cards in the show pen.

Anyone wanting to introduce a long distance family or indeed wanting a cross could do a lot worse than try some of these pigeons.

Below I have listed the positions of the two main pigeons my family is based on at the present time:

Blue hen NEHU77S1041 Champion Pickhead, flew Bourges 538 miles three consecutive years on the day. Following positions: 1st Club, 4th Fed, 27th UNC 6,357 birds; 1st Club, 1st Fed, 8th UNC 7,181 birds; 1st Club, 5th Fed, 31st UNC 5,392 birds. **Total cash winnings approximately £1,400.**

Red cock NEHU86EC3585 The Broken Keel Cock, 1st Club, 10th Fed Abbeville 327 miles; 5th Club, 16th Fed, 137th UNC Beauvais 377 miles, 11,460 birds; then flying Bourges 538 miles four consecutive years, twice on the day and twice early the next morning, 2nd Club, 10th Fed, 81st UNC 3,709 birds; 1st Club, 11th Fed, 87th UNC 2,468 birds; 1st Club, 3rd Fed, 48th UNC 2,735 birds; 1st Club, 1st Fed, 30th UNC 3,000 + birds.

Total cash winnings approximately £1,100 to date.

A limited number of youngsters for sale, prices on application.

H HADFIELD & SON
16 High Row, Loftus, Saltburn, Cleveland TS13 4SA.
Phone: 0287 643693 between 8 and 10pm.

W BOWEN 'Taffy'
172 Ystrad Road, Ystrad, Rhondda, Mid Glam, South Wales CF41 7PP
Phone: 0443 433611
1933 Founder member Welsh HPU 1993
1943 Founder member Welsh GNFC 1993

SUPER SEARCHER
1st All Wales Research Race, 1st Welsh North Road Fed, 13,065 birds.

EXPECTED
1st Welsh North Road Fed, 14,372 birds.

I have held more offices than any man alive or dead in the sport in the whole of Great Britain. I believe myself to be the last surviving member of the first Congress of the International Federation, attending the Congress in Lille, Northern France at the age of 28 in 1949. It was there I conceived the idea of a similar organisation for Britain, and sent out letters with a draft constitution which resulted in the Confederation being formed at a meeting in Gloucester. I later became its president. It was there also at the 1949 Congress that I first met Mons Henry Landercy, one of Belgium's leaders, who was, in fact, apart from pigeons one of the leaders of the Belgian resistance movement during the occupation, and he advised me where the best fanciers were to be found and their best pigeons and strains. This was of great interest to me as I only wanted the best from the best. When I went to Atwell Bros (great friends of mine, now deceased, but whenever I claim to be "Wales' best breeder" I am always mindful that the Atwell Bros were the best breeders of long distance pigeons ever, rivalled in my mind by the late Oliver Dix and his nephew Bert Bryant with the old Spangles). From Atwell I only had sons and daughters of **The Bosun** and **Mary Ann;**

Advertisement continued on next page

Continued from previous page

from Bert Bryant, sons and daughters of **Spangle Princess** and **The Combine Cock** or **Travis Red,** both old Logans; from W H George, children of **Mayfair Lad,** always the best from the best.

My first Belgians direct were gifts from the Belgian Union in recognition of my contribution to the sport, they were selected by the Belgian Union and were a blue hen, daughter of **De Prins,** best pigeon all of Belgium 1955 and in turn a grandson of **Champion De 45** best pigeon all of Belgium for the Cattrysse Bros and a black chequer hen, daughter of **Le Barcelona Crack** both from the loft of Gllrs De Smet of Waregem. These were followed by a son of **Champion Donkere,** grand-daughter of **De 45,** daughter of **Champion Witterugge,** grandson of **De 90, De 94,** when mated to sister of **De 87,** grandson **De Draaier,** all close to world famous pigeons. From Rene Boizard, grand-daughter of **Louvre II.** From Denys Bros, grand-daughter of **Oude Remi of 54** (before Denys won best pigeon ace Belgium with same blood) and a daughter of **Le Vieux Rouge** of Germain Imbrecht, this cock since founded world class family. From Devrient, grand-daughter of **Oude Karel.** Busschaert, grandchildren of **Tito & Dark Tito.** Janssens, Arendonk, grandson of **De Merckx, Vechter and Snipe** and a daughter of **The Fabulous 41.** Desmet-Matthys, from children of **Ouden Klaren.** De Baere Bros, son of Champion **Mosha Dayan.** W Clerebaut, **Le Bariole** brother to **Le Chateauroux.** Vermotte, a son of **Perpignan** and sons and daughters of Fed and National winners from Welsh fanciers who had their pigeons from me, and brought back the youngsters from the Classic winners of 1st prizes and again on the advice of Mons Landercy, pigeons from Mons Edmund Rochus of the strain Van Rhyn Kloeke, Desmet-Matthys, Janssen. All champion strains when blended and selected skillfully became the devastating **Kamikaze Sprinters.**

I claim to have **"the best pigeons in the world".** With these same pigeons I set racing records second to none in Europe, fanciers who have had them from me, not only win, but dominate their clubs and top their Feds. The first two pages of my 1991 advert is an exact copy of the 1983 advert, which said: "In the very first race of the season 1982 a 'Taffy' pigeon was 1st in the giant Welsh North Road Fed, the biggest Fed in Britain and probably the world and was the fastest pigeon of the year with over 300,000 pigeons competing". 1983 was a repeat of that fantastic performance (mathematically impossible), but **Davies & Hayward of Clydach Vale won 1st Welsh North Road Fed in the first race of 1983 with a grandson of 'Taffy' Bowen's Champion Van Rhyn Kloeke stock cock The Roc** and to prove it was no fluke, **were 1st & 2nd** in the second race. **The second** being from 'Taffy' Bowen's **Bolt** then **4th Open** in the third race with a brother to the first race winner, so again a grandson of **The Roc.**

In the same club **Tony Southwood won the Perth** race with a late bred down from **The Roc** to help win the trophy for best average all

Advertisement continued on next page

races of the giant Welsh Combine. This cock **Taffy's Pied** was second in
the Rhondda Valley Two Bird Club to Eric Davies of Tonypandy
who won with a **'Taffy' pigeon** which although only a yearling has
with her nestmate won several hundred pounds.

**Fred Edwards & son of Penywaun won 1st Heads of the Valleys
Fed.**

John Phillips of Treherbert won 1st Rhondda Valley Fed.

**Robert Oram and Ron Emblyn shared a nest pair and both won
1st South West Glam Fed.**

Stuart Garn won 1st New North Road Fed.

John Freeman won 1st Swansea Valley Fed.

These freely notified to me confirm the excellence of this family of super
racing pigeons, the above few birds mentioned beat tens of thousands
of pigeons, any of which cost a lot of money, being bred from the "best
in the world", so, if the "best pigeons in the world" finish behind the
Kamikazes, what does that make the Kamikazes?

Now down to earth, I haven't got the "best in the world", but I have
been annoyed by such claims, and at officials talking of "Valuable Studs"
when **I honestly believe I have as good as they come, and better
than the above.**

"OLD SPECKLEHEAD"

"THE LITTLE KOO"

photo by Anthony Bolton.

THE CHAMPION STOCK PAIR
THEIR PROGENY WON FIRST PRIZES IN CLUBS, SPECIALIST CLUBS, OPENS,
FEDERATIONS, WELSH NATIONALS, QUEENS CUP AWARDS ON AN UNPRECEDENTED
SCALE.
BOTH BRED AND KEPT TOGETHER FOR A DECADE BY W.BOWEN "TAFFY".

The preceding is based on my 1991 advert. Repeated hospitalisation
has made it difficult to do the writing necessary to keep up-to-date with
records etc. The origins of my pigeons may be of interest, but it is what
they are doing now that matters most to fanciers whose interest is racing
pigeons to win the very best competitions.

The above are only a fraction of the winners who have freely notified

Advertisement continued on next page

me of their outstanding winning at all levels of competition and who freely refer to me as the "Master Breeder". This made me feel self-conscious at first, but not now, for I honestly do not know of any breeder, alive or dead in Wales, to even approach breeding the winners, or more important the good pigeons that have been bred by **Taffy Bowen** creator of **The Kamikazes.** I say this fully conscious and grateful to all good fanciers and good pigeons that made all this possible.

The foregoing is a repeat of previous years' adverts, in an ongoing way, so fanciers can easily see that the outstanding successes were no "flash in the pan". These performances in Open competitions from 60 to 600 miles have been described as "out of this world", "fabulous" etc, and if listed would add a few more pages to this already long advert.

To set the above record straight the Carlisle Welsh National winner did, in fact, top the Combine so no correction was, or is due. **Grandchildren of The Roc, now known as The Golden Roc, set up performances in 1986 that have never been equalled, much less surpassed;** winning in every race from first race to the last. **In the first race of the season 1986 his four grandsons won 1st, 3rd, 7th & 8th** in a race open to all of Wales. The 1st prize-winner went on to win four 1st prizes in club racing.

Reference to Squills of 30 years ago will show I was selling pigeons at £5 each, at the same time I was selling really good tic beans at £1 per cwt. Escalating costs, and incapacity preventing me shopping around for corn, force me to ask more for these in 1993. Ask for details.

I won't use the words "cheap" or "half-priced" as this would degrade me and my pigeons, but a comparison of my own costs as the above examples makes clear an adjustment is justified. Fanciers can be assured that if a youngster is not worth two hundredweight of corn it will not be reared, and they could never be sent out.

If I am still afloat the same advert will appear with an additional piece next year. I will then try to list all the 1st Fed winners notified to me in the past decade.

I will end this as I began every advert, every year for many years past, with never a challenge. **"The past season has yet again proved Taffy Bowen to be the best breeder of winning pigeons in Wales. Taffy pigeons winning top prizes in clubs, two-bird clubs, Open races, Feds, Combine and Welsh Nationals, on a scale never equalled by any one family of pigeons, they win everywhere they go".**

W BOWEN 'Taffy'
172 Ystrad Road, Ystrad,
Rhondda, Mid Glamorgan,
South Wales CF41 7PP
Phone: 0443 433611

JOHN CARMICHAEL — KIRKPATRICK

Lerwick North — Pau South
1st, 2nd, 3rd, 3rd, 4th, 5th Sect NFC

Scotland's fancier of 1984, the late John Carmichael described his family, which was primarily Kirkpatrick: "The concordes of the sky and aristocrats of the racing pigeon world". They are birds of intelligence which has seen them through many arduous 500-600-mile races to win at all levels of competition. I have won on both north and south routes with these same pigeons.

ROMANCE EXPECTED. 1st Sect, 65th Open NFC Nantes, 10,951 birds; 17th Sect YB NFC Guernsey.

NOBLE ATTEMPT. Mealy cock. Tenth Sect, 34th Open MNFC Angouleme, 14 hours 11 mins on wing.

NOBLE ENDURANCE. Red cock. Ninth Sect, 175th Open Pau NFC.

2003. Red hen. Eleventh Sect, 477th Open Pau; 1st Club Exeter; 5th Sect NFC Nantes.

THE MULTIEGG HEN. Mealy. Second Sect, 36th Open NFC Nantes; 12th Sect YB National Guernsey; 208th Open Angouleme; flown Pau.

NOBLE LADY. Red Pied. Third Sect, 172nd Open NFC Nantes.

JUST FOR JOHN. Red Pied. Fourth Sect NFC Nantes; 7th Sect, 23rd Open MNFC Angouleme.

Since turning to South Road racing in 1981 and with the change of address which made racing impossible in 1982, my National positions with the NFC and MNFC have been: 1st, 2nd, 3rd, 3rd, 4th, 5th, 7th, 7th, 9th, 9th, 10th, 11th, 11th, 12th, 13th, 14th, 17th, 18th, 20th, 20th, 25th, 27th, 32nd, 43rd, 49th, 50th, 53rd, 54th, 59th, 60th and 80th Sect; 23rd, 34th, 57th, 62nd, 65th, 74th, 94th, 96th, 169th, 170th, 172nd, 173rd, 175th, 179th, 183rd, 186th, 208th, 232nd, 330th, 336th, 342nd, 384th, 477th, 504th, 528th, 532nd and 538th Open.

For more details see RP Pictorial May 1977, no 89, vol 8; Consistent Lerwick Flier of June 1982, no 150, vol 13; Scottish Eyes, Roundabout Success RP January 1992, Old Comrades Show Judge, RP 28 December 1984; RP Pictorial December 1988, no 228, vol 19, Kirkpatrick Special and the late John Carmichael; and Squills 1978-1992.

A few OBs will be for sale, YBs available at times thoughout the season from £25. Late breds, after OB racing, from £20 plus carriage.

Inspection by appointment only. Apply:

RICHARD HOWEY
17 South View, Kislingbury, Northampton.
Phone: Northampton (0604) 832182 or 770454 — Before 10pm please.

S & A PIGEON SUPPLIES LTD

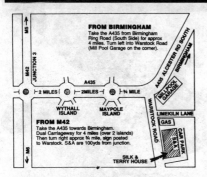

SPECIALIST SUPPLIERS TO CLUBS AND FANCIERS

Complete range of secretaries' requirements always available from stock, ie dials, rolls, rubber race rings, ink ribbons, miniature diplomas, etc.

We have a full range of race result computer systems available, usually from stock.

Send for details and sample print-outs.

FULL RANGE OF TROPHIES ALWAYS AVAILABLE.

Cups, plaques, statuettes, medals — good quality in-house engraving service.

FANCIERS — ALL YOUR BREEDING AND RACING REQUIREMENTS OUR SPECIALITY

Everything from plastic Widowhood fronts, baskets, troughs, drinkers, fountains, and many more items.

Opening hours: 9am-5pm Monday to Friday; 9am-12 noon Saturday; 10am-12 noon Sunday.
NB. Sunday opening is during racing season only.

Advertisement continued on next page

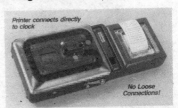

MICHAEL SPENCER

Higher Park House, Barnoldswick, Lancs
Phone: 0282 813240

National long distance winning to the Pennine Hills of North East Lancashire

SCOTTISH LADY
The Foundation of my long distance family.

1st Pendle MW Mangotsfield;

1st NEL Fed, 2nd Amal Rennes 402 miles;

1st Open NEL 2-Bird Nantes Club, 147 members, 257 birds smash, ten in race time;

1st Lancs & Yorks CC Club only bird on day, 3rd Sect L, four on winning day;

14th Open NFC Pau 734 miles, 6,928 birds, one bird sent winning £1,217;

2nd Lancs & Yorks CC Club, 12th Sect L, 81st Open Pau;

The only bird in England to appear twice in first 100 between 1982-84.

BORDER PRINCE
Grandson Scottish Lady.

1st Open National Flying Club Nantes 10,591 birds 1991 creating two National records. Furthest flying Nantes winner and most northerly loft ever to win a NFC race.

Also 1st Club 3-Bird Yearling, 10th Amal Rennes, 402 miles, 1,006 birds 1990.

The nestmate to Border Prince won in 1991 1st Club, 2nd Fed, 4th Amal Niort 529 miles; 10th Lancs Combine 1,651 birds. Convoyed and liberated with Combine not allowed to appear on result.

Advertisement continued on next page

Continued from previous page

MICHAEL SPENCER

THE SCOTTISH LADY FAMILY

Long distance National winning to the Pennine Hills of NE Lancs, *winning against all odds*, my birds fly 200 miles further than first drop and 50 miles further than first drop in Section L. With an average entry of approx four birds per race these birds won the Lancashire Rose Trophy — awarded for best average Section L Nantes, Pau and YB race — 1988, 1991, r/u 1990. No fancier has ever won this trophy to such a northerly location, indeed no one within 20 miles of me has ever won it.

Scottish Lady was bred by Hugh & John Innes of Edinburgh. I bought her as a squeaker. Here are two of the reasons why: grand-dam Kingston Treetop, 1st SNFC Avranches, 505 miles; 8th SNFC Beauvais, 498 miles; 9th SNFC Cheltenham YB, 282 miles. Great-grand-dam Kingston Again, 130th SNFC Avranches, vel 830, 3,498 birds; 106th SNFC Rennes, 5,609 birds; 58th SNFC Avranches, 4,033 birds, vel 695; 31st SNFC Rennes, 6,750 birds, vel 1262; 22nd Avranches, 4,528 birds, vel 943.

Almost all of my present long distance team are bred down from this great hen.

My results this year in Channel races are as follows:
1st Club Sartilly Barnoldswick PFC, 357 miles; 1st Open Chatburn CC Club Guernsey, 306 miles; 1st Open Lancs-Yorks CC Club Guernsey, 306 miles; 1st Club 3B Yearling Rennes Barnoldswick PFC, 402 miles; 1st Club Niort Barnoldswick HS, 529 miles; 1st Club Niort Barnoldswick PFC.

1st NEL Fed Niort, also 5th & 6th, timing three birds in 18 minutes, eight birds on day in Fed; winning Niort for fourth consecutive year, being 1st Fed, four on day; 2nd Fed, two on day; 2nd Fed, 18 on day; 1st Fed, eight on day. This hen, after winning 1st Fed Niort, vel 1126, was sent 14 days later to Saintes NFC, 568 miles, to win 4th Lancs-Yorks BCC Club, eight on day, 19th Section L, 270th Open, Vel 1222, 5,070 birds. Her dam a direct daughter of Scottish Lady x White Nose, 1st Niort at seven years of age, timed at 10.05pm, only bird on day (two in Fed) 1990; also 1st Niort Pennine 2B 1986, two on day; 1st Nantes, 461 miles as yearling.

Many other prizes have been won this year, some fifteen 1st prizes won so far despite putting some of my best racers in stock loft. Over the last 15 years these birds have won approx half of all Channel races in Barnoldswick PFC. In addition I am the only fancier in North East Lancs ever to win the three premier specialist long distance clubs where the best fanciers and birds compete. 1st Lancs-Yorks BCC Pau 1982, only bird on winning day, 4B limit; 1st NEL 2B Nantes 1981, ten birds in race time; 1st Pennine 2B Niort 1986 & 1989. Two birds, only bird on day respectively.

On every race I am furthest flying fancier.

Youngsters will be priced from £50 each.

Satisfaction guaranteed or money back.

MICHAEL SPENCER
Higher Park House, Barnoldswick, Lancs
Phone: 0282 813240

WEBSTER BROS OF ASHBOURNE

The Long Distance Specialists

The Webster strain created from the Shetland and Barcelona blood of winners from 500 miles North to 700 miles South.

During the coming season we will again have a limited number of exceptionally well bred youngsters for sale, from our well-known long distance strains. Only birds of the highest quality will be sent out. All bred from many generations of consistent long distance winners, many times on day of toss from 50-700 miles.

Weekly terms if desired and special price for orders of six youngsters or more.

Note: These birds have won many times from 500 miles and over and the only pedigree is the basket.

Wins in the strong 15-B Ashbourne FC (100 members) and 3-B Ashbourne SC (60 members).

South Road
- 1st Worcester
- 1st Gloucester
- 1st Bath
- 1st Temple Combe
- 1st Swindon
- 1st Poole
- 1st Bodmin Rd
- 1st Leamington
- 1st Weymouth
- 1st Guernsey
- 1st, 2nd, 3rd Rennes
- 1st Dol
- 1st St Malo
- 1st, 2nd, 3rd Marennes (500 mls)
- 1st, 3rd, 4th Angouleme (520 mls)
- 8th, 20th, 21st, 40th Open Pau (700 mls NFC)

Long Distance Cup won outright

North Road
- 1st Normanton
- 1st Harrogate
- 1st Leeds
- 1st Morpeth
- 1st Berwick
- 3rd, 4th Banff
- 1st, 2nd Arbroath
- 4th Thurso
- 1st, 2nd, 5th, 9th Lerwick (500 mls)

Long Distance Cup won outright, along with many others

All birds created from the old long distance Logan, Gits and Barker blood

All sires and dams of many long distance winners, many times on day of toss.

Advertisement continued on next page

Continued from previous page

Youngsters out of the nest £25 each or £23 each for six or more.

— A booking fee secures birds. —

Also a limited number of eggs £15 a pair. *Any clear eggs will be replaced with eggs or youngsters.*

All birds on five days' approval, but no responsibility for birds accepted in any way after seven days.

Stock birds and racers, late breds, at workingmen's prices, terms or cash. If you want a strain that has the stamina for flying 16-17 hours on day of toss in any weather you can do no better than to have Websters flying for you.

If you have never experienced the thrill of seeing a bird land at night on day of toss to win from the far off distance races, as we have done many times, you will be well on your way to doing so. It is a thrill that no large amount of pool or prize money won from short races can ever bring to a true fancier. For no matter the weather, storm or shine, they win and are second to none as proved many times in the past for novices and old hands alike.

All winners and sires and dams of many long distance winners, many times on day of toss.

Advertisement continued on next page

Continued from previous page

B H, ALL-A-LONE
Winner of many prizes
from 50-700 miles, 21 Pau
KO Cup longest races.
Dam of many winners.
Osman Cup winner.

B Ch, CHAMPION COCK
Winner of many races
from 60-600 miles. Sire
and grandsire of many
winners. Osman Cup
winner.

B Ch C, SUPERSONIC
Late bred. Winner of
many prizes from 50-600
miles. Sire and grandsire
of many winners.

**B C, winner at all
distances, sire of many
winners.**

**B H, winner at all
distances, dam of many
winners.**

**B C,
ASHBOURNE SUPREME**
Winner of many 1st prizes
including 8th, 20th, 21st,
40th Open Pau, 700 miles,
NFC. Sire of many
winners.

**Bk P C, winner at all
distances, sire of many
winners. Late bred.**

**B C, winner at all
distances, sire of many
winners.**

**B Ch H,
SHETLAND TWILIGHT**
Winner of two 1sts
Lerwick, 511 miles, dam
of many winners. Osman
Cup winner.

Advertisement continued on next page

**ALL WINNERS OF
500 MILES AND OVER
ON DAY OF TOSS**

**All bred and
raced by
WEBSTER
BROS**

WEBSTER BROS
Old Derby Hill
Ashbourne, Derby DE6 1BL
Phone: 0335 45517 (after 9.30pm)

VISA — ACCESS
IF BUYING BOOKS OR SUBSCRIBING FROM
OVERSEAS CONSIDER USING A CREDIT CARD.
PAYMENT BY A CHEQUE DRAWN ON A BANK
OUTSIDE THE UK ADDS ABOUT £5 TO THE
COST OF A BOOK.

PAYMENT BY INTERNATIONAL MONEY
ORDER MEANS A VISIT TO THE POST OFFICE
AND EXTRA CHARGES.

Credit cards cost us money but it is worth it because it
makes payment simpler.

**JUST SEND YOUR CARD NUMBER AND THE
EXPIRY DATE WITH YOUR ORDER.**

KIRKPATRICKS *bred and raced by*

HARRY MOORE

Chequer 4646.

Sole Survivor.

Old Faithful.

All Alone.

The above four birds between them have taken the following positions from races over 500 miles: Nine 1sts, twice only bird in race time; three times only bird on day; plus four meritorious awards.

Numerous positions by these four birds under 500 miles includes 1st Federation by one hour, etc. *Many times the loft has taken first four positions in race.*

Occasional birds for sale.

Harry Moore, ''Dorincourt'', 5 Bell Lane Fosdyke, Nr Boston, Lincs PE20 2BS
Phone: 0205 85681

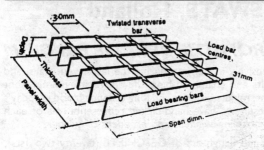
KAREL HERMANS
Super Sprinters

Since acquiring these pigeons and never having flown old birds previously, in three years these birds have taken me to the top position in a very competitive club with 25 flying members. 1991 OB results flying only 14 cocks, some of which were late breds: Stafford, 1st, 2nd, 3rd, 4th, 6th; Worcester, 1st, 2nd, 3rd, 6th, 7th, 9th; Worcester(2), 1st, 2nd, 5th, 6th, 7th, 8th, 10th; Gloucester, 1st, 2nd, 3rd, 4th, 5th; Mangotsfield, 2nd, 7th, 8th; Weymouth (1), 4th; Weymouth (2), 1st, 3rd; Gloucester (2), 2nd, 3rd, 4th; Gloucester (3), 1st, 2nd, 3rd, 5th. These birds won 47 prizes in nine races in 1992. Winning OB Av again, OB Points Cup, OB of the Year Cup, third year running.

YBs from above and stock birds £35 each with full pedigrees.

Callers welcome. All verifications available. *None sold locally.*

M Proctor. Phone: 0282 414837, after 6pm

Advertisement continued on next page

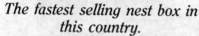

Advertisement continued on next page

Advertisement continued on next page

Advertisement continued on next page

Advertisement continued on next page

BODDY & RIDEWOOD

Bring the family for a day out at the seaside. We're only 500 yards from the beach.

VISIT OUR SHOWROOM

OPEN 9am-5.30pm MONDAY-SATURDAY

Boddy & Ridewood has, through our mail order catalogue, always offered you the best and the ease and convenience of choosing from the world's largest selection of pigeon equipment. Come and see for yourself literally tens of thousands of pigeon racing products that are now on show and in stock in our large modern showroom and grain mixtures display where we can offer over 2,000 bags of top quality Versele-Laga and other grain mixtures at very competitive prices.

Our friendly staff will be more than willing to assist or offer advice.

WE LOOK FORWARD TO SEEING YOU SOON

BODDY & RIDEWOOD
THE WORLD'S LARGEST MAIL ORDER SUPPLIER OF PIGEON RACING EQUIPMENT
Sussex Street, Scarborough, N Yorks YO11 1EH
Phone: 0723 375123 or 353332

Advertisement continued on next page

The great
JULES JANSSEN

He was known as the "Father of the Sport" in Belgium. For many years he was the Director of Finance for Belgium.

His early days were spent helping the famous Charles Wegge look after his birds. When he decided to have his own loft of birds Charles Wegge gave him the pick of his lofts.

He soon became famous for his breeding of this great strain. In his handwritten stud book (1893 onwards), now in my possession, which was given to me by the late Tommy Buck, who specialised in this strain, and bred more birds for stock which in turn produced so many great National champions, more probably than any other breeder in the British Isles.

Jules Janssen birds were used by every great fancier in Belgium, in fact for years there was not an outstanding loft in Belgium and in other countries that

The late Jules Janssen

did not contain his birds. His stud book contains the following fanciers who had birds from him:

G Gits, Jurion, Grooter, Delbar, Vandevelde. Van der Espt founded his loft on his birds (see "The Old Belgian and British Strains), Janssen of Arendonk were founded on this strain. He mentions the birds he supplied to Coopman, Carpentier, Pittevoil, Gallez, Delmotte, Vekeman, and many others. In later years Dr Bricoux, Stassart, Sion, Tremmery and Havenith, concentrated on his birds. The Cattrysse family via Vandevelde were also of this strain. N Barker had a bird enter his loft from Janssen, and Janssen made him a present of the bird, which in turn bred many good birds for him. He later let J W Logan have this red cock, which Logan considered his best breeding cock. (See Dr C R Wright's article in BHW Diary 1945.) The great Jim Kenyon's black pieds were Janssen from T Thorogood's lofts which had the Janssen family, that is why Jim Kenyon crossed his birds later with the best of Delbar's with great results. I had the last four pairs off J T Clarke of Windermere. His birds were Janssens as were the well known J Brennan's of Kilmarnock, of which I had his last six pairs, these often produce his old chocolate family. All the above men were great fanciers, at all distances, up to 800 miles, year after year. F W Marriott during the 1930s had birds from T Buck year after year, mostly Janssen x Reys.

Advertisement continued on next page

PURE JANSSENS

The **Janssens** were pure **Wegge** birds. **Jules Janssens had all of his original birds from** the great **Charles Wegge,** whom he knew very well. **In turn** the famous **Dr Bricoux founded his loft on Janssens, as did Sion, Stassart** and many other famous lofts in Europe. In my possession I have Jules Janssens' original stud book going back to 1893.

On the **death** of **Janssens, T Buck purchased** all **remaining birds in his loft.** There are **many fanciers in this country and on** the **Continent** who **considered T Buck to have** had the **finest loft** of **stock birds in Europe.**

These great birds are now settled in my lofts, and youngsters will be for sale during 1991. Every effort will be made to give fanciers the same sincere and honest service that made T Buck so highly esteemed throughout the pigeon world.

Youngsters £40 per pair
Yearlings available

A MOST COMPREHENSIVE AND ILLUMINATING STUD LIST
PROLIFICALLY ILLUSTRATED. PRICE £1.50.

Advertisement continued on next page

Advertisement continued on next page

PURE BASTINS

I have at **present ten pairs** of **Lucien Bastins, and I very** much **doubt** if there **are any purer birds** of **this strain in this country** today.

They are **from birds I purchased** from **T Buck** during 1930-35. these were from the **originals, purchased** from the late **Alf Darbyshire who imported** them **from Mons Bastin. The family** was **based** on **pure Hansennes,** and have been kept pure over the years.

They have always done very well wherever I sent them, particularly in America and South Africa.

Mostly Dark Cheqs, Bronzes, Pieds and Grizzles. A lovely type of racer.

Youngsters £40 per pair

RED CHEQ COCK
Another big winner of Fed and Combine prizes.

SUE
Winner of over £1,500 in Open races

My apologies to those disappointed customers last year. The demand was such that I could have sold many more than I was prepared to breed from this valuable stock. — Jack Lovell.

DEAN PALLATT

VOSSE WITTEKOP
The world's greatest direct Van Reet stock cock, bred by Staf van Reet, sire, grandsire and great grandsire of 1st prizewinners all over Europe, a direct son of Daniel, winner of fifty-seven 1st prizes when paired to a daughter of the Good Vosse, winner of twenty-two 1st prizes. Owned by Dean Pallatt.

GOOD VALE 83
Bred and raced by Staf van Reet and was possibly the No 1 race cock in Staf's loft at the time of purchase, a winner of fifteen 1st prizes, a direct son of Vosse Wittekop. Owned by Dean Pallatt.

GREAT VALE 81
Bred and raced by Staf van Reet, the oldest son of Daniel, has been in Staf's breeding loft for eight years, this must prove his ability as a champion breeder, a direct son of Daniel, winner of fifty-seven 1st prizes. Owned by Dean Pallatt.

Grandchildren for racing £250 for six
Grandchildren for stock £500 for six

STAF VAN REET

THE GOOD ONE 82
Bred and raced by Staf van Reet, one of the best racing sons of Dikke Prinz Good One, winner of twelve 1st prizes and is proving to be an outstanding breeder. Owned by Dean Pallatt.

THE ELF
Bred and raced by Staf van Reet and was a fantastic racer as a yearling, winner of six 1st prizes and then put in the breeding loft by Staf, a direct son of Dikke Prinz, winner of twenty-six 1st prizes. Owned by Dean Pallatt.

THE PEGGER
Bred and raced by Staf van Reet, a winner of seven 1st prizes and then placed in the breeding loft of Staf van Reet, a direct son of Dikke Prinz, winner of twenty-six 1st prizes. Dikke Prinz and Daniel are brothers. Owned by Dean Pallatt.

Grandchildren for racing £250 for six
Grandchildren for stock £500 for six

DEAN PALLATT

DONKERE WITPEN DANIEL
Bred and raced by Staf van Reet, winner of seventeen 1st prizes, also sire and grandsire of many 1st prize-winners including our own CAR WINNER 1990, a direct son of Daniel. Owned by Dean Pallatt.

BLUE WITPEN DANIEL
Bred by Staf van Reet and was one of his most famous stock birds, a true champion breeder, sire and grandsire of many 1st prize-winners, a direct son of the immortal Daniel. Owned by Dean Pallatt.

LATE WITTEKOP 86
Bred by Staf van Reet, this striking chequer pied cock was put instantly into Staf van Reet's breeding loft as a yearling where he has proved his ability to breed many winning pigeons for Staf. Wittekop 86 is a direct son of Donkere Witpen Daniel, a winner of seventeen 1st prizes. Owned by Dean Pallatt.

Grandchildren for racing £250 for six
Grandchildren for stock £500 for six

STAF VAN REET

SCHONE VALE DANIEL

Bred and raced by Staf van Reet, a winner of twelve 1st prizes. Schone Vale means beautiful mealy and believe me he is, a direct son of Daniel being a brother to Vosse Wittekop. Owned by Dean Pallatt.

SCHONE GENOPTE DANIEL

Bred and raced by Staf van Reet, a winner of four 1st prizes and then put into Staf van Reet's breeding loft as he is a direct son of The Daniel. Owned by Dean Pallatt.

THE 190

Bred and raced by Staf van Reet and was one of his best race birds at the time he came to us, a direct son of Vosse Wittekop. Owned by Dean Pallatt.

Grandchildren for racing £250 for six
Grandchildren for stock £500 for six

DEAN PALLATT

THE FAVORIET
Bred and raced by Staf van Reet and was one of his most prolific racers, a winner of twenty-two 1st prizes, a direct son of the Favoriet 77, two times ace pigeon. Owned by Dean Pallatt.

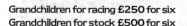

THE GOOD BLUE 83
Bred and raced by Staf van Reet. In Staf's own words, "He was always at the front with the winning pigeons". Himself a winner of eight 1st prizes and a direct son of the Vosse Wittekop. Owned by Dean Pallatt.

THE STUD
Bred by Staf van Reet, a direct son of Donkere Witpen Daniel. The Stud is sire of our CAR WINNER 1990. Owned by Dean Pallatt.

Grandchildren for racing £250 for six
Grandchildren for stock £500 for six

YOUNG TOPPER
Son of Topper x White Star, daughter of
Rapido 69.

WHITE CRUSADER
From White Arrow x Shy Girl.

SNOW KING
Son of Topper x Classic Queen.

GAY ARROW
Son of Black Arrow, son of The Little
Black.

WHITE LADY
From White Emperor x Snow Queen

Some of my White Rapido stock hens.

KING ALFRED COCK
Sire of three Combine winners + 20 winners. Grandson of the famous Legend, brother of White Emperor.

Some of my white stock birds.

GAY KING
Top son of White Emperor.

Some young weaned White Rapido Busschaerts.

YOUNG EMPEROR
From White Emperor.

WHITE PRINCESS
Top daughter of White Emperor.

Special note: My stock birds housed here include many Fed, Open and Combine winners with up to 13,000 birds.

Some trophies won in the 1991 season.

★ *SATISFACTION GUARANTEED* ★

Superb specimens. Weekly or monthly payments.

Selected matched pairs £90 (£25 deposit). Selected six £250 (£50 deposit).

Orders dealt with in strict rotation on receipt of deposit.

Confirmation letter sent with every order. *Birds sent by Interlink (£10 carriage required).*

Send for stud list — please include £2.50 towards post and packing.

Special note: All lofts are alarmed and sound monitored, and patrolled by my Dobermans.

WARREN FOSTER

Valley View Lofts, 4 Valley View, Greenhithe Kent DA9 9LU. Phone: 0322 385541

Fax: 0474 335255 (business); 0322 385541 (home)

● *Export enquiries welcome*   ● *Visitors welcomed by appointment only*

I have won in 1992 over 200 positions including forty-six 1sts, forty-eight 2nds, thirty-seven 3rds, thirty-two 4ths, twenty-eight 5ths, sixteen 6ths, etc.

I can offer YBs for sale in 1993 from £40 each or select matched breeding pairs.

Some highlights of the 1992 season — **OB Fed positions:**

May 10 — 4th, 5th, 6th, 7th Fed Tonbridge, 2,210 birds.
May 16 — 1st, 2nd, 4th, 7th, 10th, 12th Tonbridge, 2,535 birds.
May 23 — 2nd, 3rd, 6th, 9th, 10th, 23rd, 24th, 25th Eastbourne, 2,060 birds.
May 31 — 5th, 6th, 7th, 8th, 9th, 10th Clermont, 1,112 birds.
June 6 — 9th, 14th, 18th, 19th, 22nd Tonbridge, 1,351 birds.
June 13— 3rd Provins, 398 miles, 958 birds.
June 20 — 1st, 12th, 22nd, 23rd Tonbridge, 1,467 birds.
July 18 — 1st, 5th, 13th, 18th, 20th, 22nd, 23rd Clermont, 679 birds.

YB Fed positions:

August 1 — 4th, 19th, 23rd Peterborough, 2,580 birds.
August 8 — 3rd, 4th, 5th Huntingdon, 2,577 birds.
August 15 — 9th Huntingdon, 1,459 birds.
August 22 — 1st Stevenage.
August 29 — 14th, 21st, 22nd, 24th Tonbridge, 1,293 birds.
September 26 — 1st, 2nd, 4th, 6th, 8th, 11th, 15th, 20th Stevenage, 998 birds.
October 4 — 1st, 2nd, 3rd, 4th, 5th, 6th, 7th, 9th, 10th Glanford Open Stevenage.

HOW MANY OF TODAY'S STUDS CAN PRESENT RESULTS SUCH AS THESE?

Many averages, trophies, etc, won once again and over 60 prizes in Scunthorpe Fed, up to 2,747 birds per week!

EXCITING NEWS FROM THE AXHOLME STUD

After a number of visits to the Continent we are delighted to announce that we have imported more than 50 pairs of the brilliant **Robert Venus** super sprinters and all distance birds — all direct children of his world-famous breeders and racers.

These birds were personally selected by **Ian Axe** and they really are out of this world! We now own direct children of champions Olympiade 1st and 2nd National, ten 1sts, Olympiade Gold Medal etc; Lord seven 1sts, son of Stamvater I eighteen 1sts etc; Brilliant seven 1sts including

Advertisement continued on next page

Continued from previous page

My UK Agent is
IAN AXE
The Hall, Owston Ferry,
Nr Doncaster, S Yorks DN9 1AW.

Ian is my sole UK agent and owns the finest collection of my birds outside of my own loft. My UK stud book is available from Ian Axe, phone 0427 72327.

The legendary Supercrack, responsible for a dynasty of National and International winners.

Contact:
ROBERT VENUS
Kwellemolenstraat 19, Oeren
Alveringem, Belgium
Phone: 010 32 58 28 92 43

Advertisement continued on next page

TROPHIES WON IN 1991

Just as loft space is important so is advertising space. So read the Editor's introduction in last year's Squills 1992, pages 3 & 4. Having read it, digest it; then read pages 356 & 357 (1992).

Having had another great season in 1992 pictured are just four of five Open race winners in 1992. All hens?

FULWOOD GOLDMINE — 1st Open Leyland Midweek FC Mangotsfield, 128 birds, winning £242. Winner of numerous prizes and dam and grand-dam of numerous winners including Fulwood Mint, etc.

FULWOOD MINT — 1st Sect C, 4th Open NW Classic Club Rennes, 188/958 birds, winning £274. Winner of numerous prizes and dam of numerous winners including meritorious award winner Fulwood Leading Lady, Fulwood Countess, Fulwood Duchess and Fulwood Girl, etc.

Fulwood Lady Express — 1st Open Preston 2B Weymouth YB, 62/117, winning £225, vel 1278. Beating next Fed bird by 50 ypm, approx 12 mins. She also beat a further 64 birds liberated at the same time in the Golden Ring race. Bred for me by Mr & Mrs R Kirkbride.

FULWOOD GIRL — 1st Open Lostock Hall FC, Shrewsbury OB, 50/302, winning £113.

My cocks also win — enough said.
Selected YBs for sale in 1993.

JOSEPH DORNING
The Conifers, 138 Lightfoot Lane
Fulwood, Preston, Lancs PR4 0AE
Phone: 0772 861478

Advertisement continued on next page

Advertisement continued

EYESIGN COCK

EYESIGN HEN

GERARD

THE 21

THE 17

DAPPLE BLUE

All the above six birds are Pure Vandeveldes.

This family is the purest in England today, countless winners from them.

Advertisement continued on next page

SOME OF THE TROPHIES WON BY JOHN CROWDER

It is impossible to estimate how many winners have been bred from my stock, they are winning all over England. Visitors state they are the finest ever seen. A list of just a fraction of those that win with my birds: M Bower, Brook & Crookes, Mr & Mrs Meyrick, D Pyle, R Hudson, Robinson & Son, Pepper & Lender, Hovell & Roberts, R Cooper & Son, Costello & Caudon, P Buxton, Steers & Baker, Gregory & Son, J Staniforth, McGlesson, Allwood & Freeman, S Moss, B Williams, Hind Bros, Lacey & Son, Richardson & Tate, A Lazel, E Linney, R Harris, R Blacknell, C Winthrop, Armstrong & Craig, W Scotland, Finkill & Dtr, Kirkham & Kirkham, Ian Beaumont, B Ancliffe, Bob Bell, N Cowton, Wilson Hurlford, F Nicholson. **Ayres & Fletcher,** flying my Vandeveldes never looked back — just a few of their wins: two longest races 1985, 1st Thurso, 1st Lerwick; 1986, 1st Thurso, 1st Lerwick; 1986: twelve 1sts, eleven 2nds, nine 3rds, ten 4ths. **Teddy Kobylanski,** my birds have won every year since he had them in 1979. I bred his 1st Gold Medal winner. A son of my special cock bred his second 1st Gold Medal winner and for good measure I bred his 1986 Thurso winner from Klare. **Mr & Mrs Ashley** started in the sport three years ago, they just dominated the North Road, top winners every year. Just for good measure they won Silver Medal with my Pluto Red family.

Advertisement continued on next page

THE PLUTO BUSSCHAERTS

MOSBORO SUPREME

MOSBORO BEAUTY

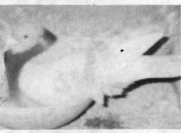

ROSQ SUPREME

MOSBORO PRODUCTIVE GENT

Note: At stock **60 direct grandchildren off Pluto; 50 direct grandchildren off Rapido; 20 direct grandchildren off Little Black.**

The foundation, champion Cover Girl, bred by Janssen Bros. Her mate for 1993 The Little Merkx, all Janssens.

1993 babies line-bred to champion Cover Girl. Price on request. My Janssens 100% pure. *Guaranteed absolute.* **They are beauties.**

Congratulations to P Rawson, winning motor car through Thor.

Continued from previous page

The No 1 home of the
WHITE PIED RAPIDOS
in England today
None bred as well as these to be had.
Three direct off Rapido.

THE RAPIDO **THE RAPIDO**

Congratulations on Brian Hovell & Roberts winning Thurso and clock, timing eight on day when very few birds homed; Mr Denny winning 1st Angouleme, 521 miles; B Antcliffe winning Nevers, only bird in race time over 500 miles. Above results all won on same day.

BABY RAPIDOS

Brian Williams wins every Channel race in one season — 12 years at the top — with my birds.

TOMMY LITTLE
WELCOME HOME LOFTS
No 1 STOCK PAIR

AIR FRANCE
Dordin-Ko Nipius. Winner of twenty-one 1sts over Channel.

KLEINE MARGE
Busschaert. 1st Fed, 22nd WDA Combine 9,099 birds.

No 2 STOCK PAIR

DREAM BOY
Busschaert. From Champion Studtopper and Champion Ireland's Dream Girl.

SYMPHONY
Busschaert. From Tot Douglas & son and Gadaffi (FWS Hall).

No 3 STOCK PAIR

GEORDIE
Busschaert. From Champion Studtopper and Champion Smokie.

GEORGINA
Busschaert. From Champion Studtopper and Bessie.

Advertisement continued on next page

Continued from previous page

TOMMY LITTLE
WELCOME HOME LOFTS

No 4 STOCK PAIR

BARCELONA BOY
Busschaert. From Champión Barcelona and daughter of Young Musketeer and Champion Shy Lass.

DAPHINE
Busschaert. From Coppi II and The Musketeer Hen.

No 5 STOCK PAIR

DEVON LAD
Busschaert. From Geordie and Sheba.

SHEILA
Busschaert. Grandparents Champion Barney, Smart Bird, Champion Young Vinnie and Starmaker.

Advertisement continued on next page

Continued from previous page

No 6 STOCK PAIR

LAWRENCE
Busschaert. From Geordie and The Larkins Hen.

PETITE FLEUR
Busschaert. Grandparents The D'Artagnan Cock, The De Koe Hen, Rambo, The Coppi Hen.

No 7 STOCK PAIR

Georgie, Busschaert from George Corbett parents; and **Penelope,** Busschaert, grand-daughter of The Gold Mine Pair.

No 8 STOCK PAIR

Geoff, Busschaert, inbred to Studtopper; and **Carol,** NEHU91DM5113.

No 9 STOCK PAIR

Coppi Lad, Busschaert from Carbon Coppi and Dusty Maid; and **Caroline,** Busschaert from son of Dark Princess and Cara.

No 10 STOCK PAIR

Alan Boy, Busschaert, bred by Alan McIlquham; and **Star Lite.**

Only three rounds bred — No late breds. Prices on application.

TOMMY LITTLE
WELCOME HOME LOFTS
6 Burnopfield Gardens, West Benwell, Newcastle on Tyne NE15 7DN.
Phone: Tyneside (091) 274 7534 or 274 4099.

THE ASHDOWN LOFTS

A R HILL (St Just)
Famous Grizzle family

J KIRKPATRICK
Bricoux-Logan- based family

Blended over many years to produce long distance racing pigeons, second to none.

My late friend, the great Arthur Hill, allowed me the pick of his loft in the years prior to his untimely death. We have concentrated on maintaining the line breeding to re-produce ãs near to the original as possible — introducing only proven quality stock of champions closely related to these immortal bloodlines, names such as:

WATCH IT
Bred and raced by the late John Carmichael, owned by Keith Wilkins, Lewes, Sussex. Fourth Sect, 12th Open SNFC, from Nantes. Grand-daughter of Lauderdale Supreme, 33rd Open SNFC Rennes.

Dutch Express, Thirty-Three, Stitchy, The Dusseldorf Cock, Cornish Lad and Lionheart are the foundation of our Hill Grizzle family.

Those of Galabank 235, Galabank 487, Duke, Prince, King — 442, 443, 1276, The Blue Annan Cock, Champion Breeder 219 and many more at the base bloodlines of the late Scottish National Ace John Kirkpatrick.

TWEEDSIDE JOHN C
Bred and raced by W Murray & Son, owned by Keith Wilkins of Lewes, Sussex. Winner of many club and Fed positions including 21st Open SMW Spec Club, 8th Sect, 8th Open SNFC Dorchester, 6,202 birds. Flown Channel five times into Scotland.

VISITORS ONLY BY APPOINTMENT PLEASE

G W KIRKLAND

Coalport House, Coalport Telford, Shropshire TF8 7HZ

Phone: Telford 580525

The Shropshire home of record-breaking National winners.

PEST — double Classic winner, 1st British Barcelona Club Nantes 1987 and 1st Midland National FC Angouleme 1989.

Nine times 1st Open in National and Classic races, three 2nds, four 3rds, and so many positions I have lost count.

Once again a very good year in 1992, being in the money in every National race, old and young. Nine races in all. 1991 1st Sect, 1st Open Fougeres National MNFC and new motor car; 1st Sect, 18th Open Bordeaux 545 miles on day NFC and finishing off with 3rd Sect, 3rd Open (provisional result) Picauville YB National MNFC. 1990: 2nd, 5th & 7th Open Angouleme National MNFC. This makes it four times 1st and one 2nd in the last six years in this race. I am only interested in National races and use club races as preparation but I still win many club and Fed races every year although I never bother to quote them.

My family of pigeons is composed of several strains, always pairing best to best and not just paper pedigrees. The base strains are **HVR, Desmet-Matthys and allied strains with Desmet-Matthys** base, ie **Verheye, De Baere** etc. I have crossed these with **Alfons Bauwens** and **Santens Bros,** with great success.

I also have a Dutch long distance family from **Chris van de Pol** from Goirle, and these have won right away for me at all distances to 495 miles.

I have won 20 RPRA awards for performances over 300 and 450 miles in Western Region RPRA.

For other performances see Year Book or National results any year — they are very much the same.

GEOFF HUNT & SON

DUSKY DIAMOND

1st British International Championship Pau, 554 miles, 1985. Flown Pau five times and Perpignan, 600 miles, twice.

EUREKA

1st British Barcelona Club Perpignan, 600 miles, 1982. Her grand-dam on both sides is the dam of Chiquita.

HERMOSA

6th Marseilles, 586 miles, 1984. 2nd Pau, 554 miles, 1985. Both with the British International Championship Club.

CHIQUITA

2nd Marseilles, 586 miles, 1985. Barcelona twice, Marseilles twice, Perpignan, Palamos and Dax. Chiquita's sire is grand sire of Hermosa.

Advertisement continued on next page

These pages of photographs are to record the performances of our family of pigeons rather than to sell large numbers of squeakers. They also give an indication as to how they are bred.

Like a pedigree in photos.

NEVER BEND

A grandson of Dusky Diamond and Hermosa on the sire's side.

A grandson of Kefaah on the dam's side.

1st Barcelona, 684 miles, 1992 BICC.

SIPSI FACH (Little Gypsy)

A granddaughter of Dusky Diamond and Hermosa on the sire's side. Same sire as Never Bend.

A granddaughter of Eureka on the dam's side.

1st Sect, 7th Open Dax with the London & SECC, 535 miles, and three weeks later 17th Dax, 534 miles, with the BICC, 1991. In June 1992 she was 5th Pau International and six weeks later won 3rd Open, 2nd BICC Perpignan, 600 miles.

FACTS OF TWELVE YEARS RACING WITH PHOTOS AND DETAILS

FAIR TRIAL

3rd Pau British International Championship Club. His dam is Hermosa, a grand-daughter of Amigo, 2nd Barcelona.

THE ROOK

Always stock. He is sire, grandsire and great grandsire of many birds to score from Barcelona. He is grandsire of The Quiet Man, his sire was Bartholomew, his dam La Oscura.

HOIST THE FLAG

1st Pau British International Championship Club. Her sire was from Bartholomew.

VICHTE PRINCESS

1st Dax British International Championship Club. From a son of The Beer x daughter of Napoleon.

We are not interested in points, averages, or championships, just taking each race as it comes with the exception of High Bid, all the photos are of pigeons that have been timed on the winning day into the British Isles.

For some more performances see page 385 Squills (1981), 380 (1982), 295 (1983), 374 (1984), 201 (1985), 224 (1986), 182 (1987), 386 (1988), 288 (1989), 304 (1990), 292 (1991), 312 (1992).

Advertisement continued on next page

THEY FLY JUST AS WELL FROM EARLY LIBERATIONS
(timing on the day) AS THEY DO FOR TWO-DAY RACES

THE QUIET MAN

1st Welsh National Pau by 1 hour 49 minutes, 589 miles for Colin Lloyd, Griffiths & son. His sire was from The Rook, and his dam was from The Rook's sister.

HIGH BID

2nd Dax British International Championship Club. His sire is Gallito x daughter of Dusky Diamond x dam of Hoist The Flag.

THE AMSTERDAM COCK

1st Amsterdam British Berlin FC for Bill Dray of Ramsgate.

RED FUTUNE

1st Libourne, 1st Nevers EECC. Sire of our 39th Open Palomos, his dam was also dam of Hoist The Flag, his sire was a son of Bartholomew.

We consider the blood of The Amsterdam Cock was one of the key introductions to our John Bank's-based (Funny Colours) family mainly through Sockeye x Sally blood.

Truman-Dicken of Wells, Somerset, also used the same blood, ie a son of The Amsterdam Cock, which helped in producing Cambridge, the grandsire of Amigo who in turn is grandsire of Hermosa. Truman, also used a John Bank's bird, a a grizzle cock he called Flash, a winner for Sam Bailey of Glastonbury. Flash from a brother x dam mating to Sally, 3rd Mirande NFC.

Advertisement continued on next page

BRIAN HOVELL & J ROBERTS

CATTRYSSE ★ BUSSCHAERT
VAN WILDEMEERSCH

ALSO OUR LONG DISTANCE FAMILY
£26,900 WON IN THE LAST 20 YEARS
£4,400 WON IN 1992

1992: NRCC Lerwick, 16th Open, 4,450 birds, with pencil blue pied hen. She won the special diploma for three times in the first 100 Open NRCC – from Lerwick 1991, 29th Open, 4,235 birds; 1990, 26th Open, 3,728 birds; 1991, 1st Sect, 2nd Open NRCC Lerwick 4,235 birds with our blue hen Natalie, grand-daughter of Valerie.

1992
OB & YB
& Scottish
Av Sutton
Central
Club

1992
OB &
Combine
Av
Mansfield
NR Club

Notts & Derby Border Fed Av 1991 and 1990
Sutton Central Club sends up to 650 birds a week.

1992 results: nine 1sts in Clubs. NRCC Perth results: 23rd Sect, 26th Open, 1,286 birds. NRCC Lerwick; 16th Open, 4,450 birds, a club record. **NRCC Thurso,** 3rd Sect, 6th Open; 7th Sect, 11th Open; 13th Sect, 24th Open; 32nd Sect, 60th Open; 45th Sect, 85th Open; 46th Sect, 90th Open; plus a quartz clock, from 4,600 birds. **NRCC Berwick YB race,** 26th Sect, 47th Open; 37th Sect, 72nd; 47th Sect, 87th Open, from 3,762 birds; winnings over £2,000 in the NRCC in 1992; also Best Single Bird Av from Perth & Lerwick.

1992: 1st, 2nd, 3rd Sect, 3rd, 8th, 14th Open **Notts NR Fed Thurso,** 1,897 birds; 2nd Sect, 8th Open **Lerwick Notts NR Fed,** 905 birds.

1992 in the **Notts & Derby Border Fed** in the first 20 in results: 1st Fed Durham, 1,391 birds; 5th Fed Northallerton, 2,813 birds; 5th Fed Lerwick, 538 birds; 8th Fed Berwick, 1,367 birds; 9th Fed Durham, 1,391 birds; 10th Durham, 1,383 birds; 11th Thurso, 1,491 birds.

Also £600 won in the **Midlands Championship Club 1992.**

Advertisement continued on next page

Continued from previous page

1991 results: 1st Sect, 2nd Open **NRCC Lerwick,** 4,235 birds; 1st Club, 1st Fed Lerwick plus Gold Medal with Natalie, grand-daughter of Valerie; 1st Sect, 2nd Open **Faroes,** 646 miles; 1st Sect, 2nd Open **Elgin YB,** 323 miles.

1991 Old Bird Av, Comb Av, Scottish Av and YB Av in **Sutton Club.**

1991 NRCC Lerwick results: 1st Sect, 2nd Open plus 29th, 39th, 62nd, 73rd, 125th Open.

1991 Fed results in the first 20: 1st Fed Lerwick; 2nd Fed Ripon; 3rd Fed Northallerton, 382 birds; 6th Fed Perth, 1,197 birds; 17th Fed Morpeth, 1,713 birds; 7th Fed Lerwick, 656 birds.

1990 results: 1st Club, 4th Fed **Lerwick,** 666 birds; 11th Open **NRCC Lerwick,** 3,728 birds; 1st Fed plus Gold Medal **Perth,** 2,333 birds. In 1990 we won five club 1sts plus 25 other prizes.

1990 Fed results in first 20: 1st Perth, 2,333 birds; 3rd, 4th, 5th **Perth Championship race,** 1,637 birds; 4th, 11th, 20th Fed **Lerwick Championship race**; 5th Fed **Thurso Championship race**; Fraserburgh 5th, 7th, 1,528 birds; Northallerton 8th, 4,534 birds; Harrogate, 17th, 4,771 birds; Northallerton, 19th, 3,534 birds. **NRCC Lerwick:** 11th, 26th 49th, 50th Open, 3,728 birds.**1989 results:** 1st Club, 1st Fed plus Gold Medal **Perth** with our good cock Tippy, 1,803 birds; 1st Club, 1st Fed **Morpeth,** 3,149 birds; 1st Club, 11th Fed **Thurso,** 1,640 birds, 15 prizes in the first 20 in Fed, eight 1sts plus two 1sts Fed; seven birds in the first 100 Open in the **NRCC Berwick YB** race, 3,810 birds.

1988 results: OB Av, Scottish Av, Comb Av in **Sutton Club;** 1st **Lerwick** Sutton Club, 15th Fed; 1st **Thurso,** 11th Fed, 1,567 birds, **only bird in club on the day,** 200 birds sent in club, 1st Club, 1st Fed **Harrogate,** 3,697 birds.

1987, 1st **Lerwick,** 10th Fed, 486 birds; 1986, 1st Club, 1st Fed plus Gold Medal **Thurso.** 1st Club Lerwick; 6th Notts Fed with Valerie.

Winners at Lerwick 1986-87-88-89-90-91-92.

A few results from our birds: Mr Wragg, Youlgrave, eight 1sts. Mr Harwood, West Leeds, one hen bred by us won eight 1sts plus 1st Fed. She bred birds to win nine 1sts in one season. Mr Ford, Blackwood, Gwent, sent me a letter with two pages full of winning pigeons from our Cattrysse birds winning up to Thurso. Mr Ling of London had one of our Cattrysse·in 1989 that won in 1990 1st Berwick Yearling race, 2nd Open London Combine, 10,000 pigeons. Mr Annetts of Birdholm won lots of good races with our birds, only bird on day Nantes, nestmate won 2nd Saintes. Brian Hyde of Morton, Lincolnshire, won 1st Peterborough Woodstone Open race from Northallerton plus £100 prize money.

Special YBs from pairs both half-brother half-sister from Valerie, 1st Lerwick, 2nd Open Faroes, 646 miles, and our long distance family £30 each.

All other YBs £20 each, six for £100, plus Interlink, plus £10 deposit.

Make cheque payable to:

Brian Hovell, 15 Garden Lane
Sutton-in-Ashfield, Notts NG17 4LF
Phone: 0623 511778.

SQUILLS DIRECTORY

Fanciers are particularly asked to include the post code in their address.

Akester D, Stepney Bungalow, Grove Hill, Beverley HU17 7QJ. Phone: 861214.
Allen Gordon, 22 Hampton Drive, Hampton Court Village, Belfast 7, N Ireland.
   Phone: 0232 647595.
Angrove J C, 66 Mongleath Avenue, Falmouth.
Ashcroft Barry, The Idle Barber, 203 Queens Road, Bradford, West Yorkshire BD2 4BT.
   Phone: 639734.
Atherton & Son, 10 Phillimore Street, Lees, Oldham, Lancs.
Atkinson, P J, 17 Cliffe Road, Gribblestone, Wakefield, West Yorks WF4 3EQ.
   Phone: 0924 258412
Axe Ian, The Hall, Owston Ferry, Doncaster, Yorks. Phone: O Ferry 327.

Bailey Mr & Mrs, 46 Vicarage Road, Harlow, Essex CM20 3HP. Phone: 02796 38155.
Ball & Sons, 7 Spring Cottages, The Rock, Ketley, Telford, Shropshire.
Bamford J, Dorweeke Cross, Butterleigh, Culompton, Devon. Phone: Bickleigh 453.
Bancroft Mr & Mrs, 20 Hoppett Lane, Droylsden, Tameside M35 7HX.
Barker & Son Mr & Mrs, 1 James Way, Immingham, South Humberside DN40 1PB.
   Phone: 0469 73500.
Barnes John, 5 Beech Grove, Barnoldswick, Nr Colne, Lancs BB8 6AE. Phone: 0282 813743.
Barr L O, Mount Juliet, Camanee, Templepatrick, Co Antrim, NI. Phone: Templepatrick 419.
Bartlett A H, Ashlea, 32 Beachwood Avenue, Frome, Somerset. Phone: Frome 3054.
Bates P J, 11 Ermin Place, Cirencester, Glos.
Bath G & M, 99 Beaconsfield Drive, Burton, Stoke-on-Trent. Phone: 0782 324823.
Beard Mr & Mrs S, 16 Market Oak Lane, Bennetts End, Hemel Hempstead, Herts.
   Phone: 0442 56528.
Bessant D, 34 Gorleston Road, Branksome, Poole, Dorset DH1 2WW.
   Phone: Bournemouth 762039.
Betts R, 29 Station Road, Shotton Colliery, Durham DH6 2JL.
Bickerdike A W & M & Sons, 'Windy Ridge', Ripon Roazd, Killinghall, Harrogate.
   Phone: 61481.
Bissett Mr & Mrs Ron, 2 Crest House, Church Road, Ashford, Middlesex TW15 2NH.
   Phone: 07842 42085.
Bloomer G B, 117 Knowle Hill Road, Netherton, Dudley. Phone: Dudley 50786.
Blowe Brian & Jayne, 83 Gorse Crescent, Stretford, Manchester M32 0UP.
Bowen W, 172 Ystrad Road, Ystrad, Rhondda, Mid Glamorgan, S Wales CF41 7PP.
   Phone: 0443 433611.
Boyd J, 6 King Edward Road, Leiston, Suffolk.
Boyd J R, 151 Cavendish Street, Ipswich, Suffolk.
Boylin Mr & Mrs R D, Blackthorn Cottage, Tyringham, Newport Pagnell, Bucks MK16 9ES.
Bradshaw Bros V, 15 Ty Fry Road, Rumney, Cardiff.
Brickley A R, 17 Hawthorne Road, Wimblebury, Cannock. Phone: 0543 75485.
Brighten Mr & Mrs C E, Woodside, Burnham Green Road, Bulls Green, Datchworth, Herts.
   Phone: 043 879 409.
Bumby G, Rose Cottage, Beadlow, Nawton, Yorks. Phone: Kirkdale 205.
Bumby & Partners, Ash Tree Dairy Farm, Top Cliffe, Thirsk YO7 3HW. Phone: 0845 577358.
Butler & Garrick, 7 Pamela Road, Immingham, South Humberside DN4 OB6.
Burke Bernard, 30 Radcliffe Avenue, Swain House Road, Bradford 2, W Yorks.

Callan T, 35 Skelton Road, Brotton, Saltburn, Cleveland.
Card R, 58 Clitterhouse Road, Cricklewood, London NW2. Phone: 01-458 6898.
Carne A, Cornhill Farm, St Blazey, Par, Cornwall.
Campion Bryan, 49 Red Lake, Ketley, Telford, Shropshire.
Chambers Jackie & Son, 8 Cloughrea Bungalows, Besbrook, Co Down, N Ireland.
   Phone: Bessbrook 830486.
Chilton Mr & Mrs T G, 3 Willans Buildings, Sherburn Road, Gilesgate, Co Durham.
Claughton A, 89 Lees Road, Oldham, Lancs. Phone: 061-624 8069.
Clarkson R, 127 Market Street, South Normanton, Derbyshire DE55 2AA.
   Phone: Ripley 810949.
Collison A, 2 Hillside Crescent, Whittle le Woods, Chorley. Phone: 74928.
Colven John, 48 Easter Drylaw Drive, Edinburgh 4, Scotland. Phone: Dean 4015.
Cook P, 10 Parkstone Road, Ropley, Nr Arlesford, Hants. Phone: Ropley 3276.
Cook Mr & Mrs W, 40 Montague St, Abertillery, Gwent NP3 1PD. Phone: 0495 216400.

Cooke B, Sunningdale, Occupation Lane, Kirby in Ashfield, Notts.
Cooke Mr & Mrs P R, 1 Lower Woodcott Cottages, Whitchurch, Hants RG28 7QA.
    Phone: Highclere 254269.
Cooper G, Duivencote, High Street, High Littleton, Nr Bristol, Avon.
Cottle Mr & Mrs P Baker, 6 Llanarth Road, Treowen, Newbridge, Gwent NP1 4EA.
    Phone: 0495 244521.
Cowell Mr & Mrs R J & M, 10 The Laurels, Gillingham, Dorset SP8 4RS. Phone: 07476 4102.
Cowood N, 130 Main Street, Goldthorpe, Rotherham. Phone: Rotherham 815798.
Cox Les & Son, Hildene, 56 Elm Grove, Bideford, Devon.
Cracknell John, 9 Granta Way, Gorleston, Great Yarmouth, Norfolk.
Cranstoun Doug, 12 Stoddart Road, Penge, London SE20.
Cullen A J, 42 Coombfield Drive, Darenth, Kent DA2 7LQ.
Cushley Mr & Mrs J, 1 Park Place, Strathblane, Glasgow G63 9DW. Phone: 0360 70257.

Davies D, Maesteg Park HS, 7 Brynles Park Estate, Maesteg, Mid Glamorgan CF34 9NH.
    Phone: 0656 736159.
Davies Williams & Davies, 40 Pritchards Terrace, Phillipstown, New Tredegar, Gwent.
Davis Fred, 20 Turning Lane, Scarisbrick, Southport, Merseyside PR8 5HY.
Dean A, 48 Heathgap, Road, Blackfords, Cannock, Staffs.
Denton Harold, Pigeon Photographer, 112 or 114 Balfour Road, Darnall, Sheffield S9 4RY.
    Phone: 0742 444947.
Dixon H, 62 Boulevard Avenue, Grimsby, South Humberside DN31 2JQ.
Dixon M, 1 Williams Court, Beckwithshaw, Harrogate, N Yorks. Phone: 501736.
Douthwaite T E, 6 Windmill Drive, Wadworth, Doncaster, S Yorks DN11 9BX. Phone: 855297.
Dudhill J, 50 Cemetery Road, Mexborough, S Yorks SL4 9PN. Phone: 588355.
Duncan W S, Bruce Cottage, Emsdorf Street, Lundin Links, Fife, Scotland.
Dunne & Co, 54 Darragh Park, Wicklow, Eire.
Dursley Ron, Bay View Stores, Mustards Road, Isle of Sheppey. Phone: 079581 256.
Dyson L A & Son, 66 Powke Lane, Rowley Regis, Warley, West Midlands.
Dyson Mr & Mrs T, 26 Badger Place, Woodhouse, Sheffield 13.

Ebsworthy D, 59 Brentford Avenue, Whitleigh, Plymouth, Devon PL5 4HB.
    Phone: 0752 776014.
Erdington P, Jesmond Dene, 116 Hesland Road, Chesterfield, Derbyshire.
Erwin Billy, Rathlin Drive, Ballmoney Road, Ballymena, Co Antrim, NI. Phone: Ballymena 41293.
Evans Edwin, 7 Hilton Lane, Great Wyrley, NR Walsall, Staffs.

Feldman R, 37 Stoneleigh Road, Clayhall, Essex IG5 0JB. (Ronfeld) 'To The Winners'.
Fisher R, 18 Ostend Court, Kemsley, Sittingbourne, Kent ME10 2TJ.
Foot J, Hyluve, Crown Road, Marnhull, Sturminster Newton, Dorset DT10 1LN.
    Phone: Marnhull 252.
Fox J D. 9 Redhill Cottages, Ludford, Lincs. Phone: Burgh-on-Bain 664.
Fox R, 134 North Street, Holmgate, Nr Chesterfield, Derbyshire.
Francis Bros & Son, 8 Greenfield Street, Maesteg, Glamorgan. Phone: 737687.
Frith V, 30 Clifton Road, London SE25.
Frosts & Son, 17 Johnson Estate, Wheatley Hill, Durham. Phone: 821553.
Fysh Mr & Mrs N H, 154 Hill Lane, Southampton, Hants. Phone: 0703 29365.
    Gibbon W, 26 Greenbank Drive, South Hylton, Sunderland, Tyne & Wear.
Giles J C, 8 London Road, Standford Rivers, Ongar, Essex.
Gill M, Ox-Heys Farm, Shelf. Phone: Bradford 677704.
Gilmore & Beattie, 18 Breda Drive, Belfast. Phone: 64320.
Godley Ken, 4 Buxton Crescent, Turf Hill, Rochdale, Lancs OL16 4TU. Phone: 357051.
Goldfinch Mr & Mrs, 39 Colebrook Road, Swindon, Wilts.
Goldsmith D F, 87 Verona Drive, Surbiton, Surrey. Phone: 01-397 1363.
Graham E & R, 23 Circular Drive, Chester CH4 8LU. Phone: 0244 674912.
Grantham W E, 290 Whersted Road, Ipswich. Phone: 50133.
Griffin F E, Jackson's Farm, Chippenham, Wilts. Phone: Kington Langley 271.
Groom F & M, 15 King Street, South Normanton, Derbyshire DE55 2AH. Phone: Ripley 813718.
Gunn D R & J P 225 Oakdene Road, Orpington, Kent. Phone: 0689 39802.
Guy D J, 9 Seaview, Sudbrook, Newport, Gwent NP6 4SU. Phone: 0291 422613.

Hall F W S, 73 London Road, Enfield, Middlesex. Phone: 081-363 2125.
Hallam Mr & Mr R K, 52 Victoria Street, Belevedere, Kent DA17 5LN. Phone: 03224 35040.
Hallam Mr & Mrs T N, 134 Bambury Street, Adderley Green, Longton, Stoke-on-Trent.
    Phone: 31009 (Delbars).
Hamilton A G, 30 Hanscombe End Road, Shillington, Hitchin, Herts.

Hamilton Bros, 14 Colister Gardens, Darnall, Sheffield S9 4HA. Phone: Sheffield 445028.
Hand Mick & Bert, 65 Stanley Gardens Road, Teddington, Middlesex TW11 8SY.
        Phone: 081-977 0864.
Handy C, 66 Lullingstone Avenue, Swanley, Kent.
Hardy F G, Fig Street House, Oak Lane, Sevenoaks, Kent TN13 1UA. Phone: 52965.
Hardy P T, 34 Mill Place, Crayford, Kent. Phone: Crayford 26401.
Hart H, 232 Green Lane, Leigh, Lancashire. Phone: 0942 606030.
Harris H, 21 Richmond Caravan Park, Richmond, North Yorkshire. Phone: 0748 4430.
Harris W H, 41 Downs Road, Slough, Berks. Phone: 43304.
Harvey W M 10 Queens Way, Hayle, Cornwall. Phone: 754460.
Hauber E E, Broadacres, Keresforth Hall, Drive, Kingstone, Barnsley, S Yorks.
Hawkins & Bennett, 6 Clarence Street, Newport, Gwent NPT 2DA.
Hayward J, 513 Pearse Villas, Dunleoghaire, 6 Dublin. Phone: 012 850 482.
Healy & Conaghy, Ballymakenny Road, Drogheda, Co Louth, Eire. Phone: 041-51364.
Heaney E & Son, 322 Skegoneill Avenue, Belfast BT15 3JX. Phone: 0232 370956.
Helm & Knox, 89 Moorlands, Prudhoe, Northumberland. Phone: Prudhoe 32647.
Henderson Alex, 20 Moss Lane, Sale, Cheshire M33 1GD. Phone: 061-969 1753.
Hogan Graham & Phil, Chez Nous Lofts, 'Urmiston', 18 Merafield Road, Plympton,
        Devon PL7 3SJ. Phone: 0752 342723.
Holbard Mr & Mrs F, 56 Endeavour Road, Cheshunt, Herts.
Holland Roy, 514 Nelstrop Road North, Levenshulme, Manchester M19 3JL.
        Phone: 061-224 6147.
Hough F & M, 26 Clarence Rd, Bollington, Macclesfield, Cheshire SK10 5LD.
        Phone: 0625 574807.
Hubbard Allan & Robinson Ian, The Nook, Holton St Mary. Phone: Gt Wenham 310962.
Hulse P & Son, 57 Walnut Drive, Winsford, Cheshire. Phone: 4835.
Humphries D A, Station House, Tipton St John, Sidmouth, Devon EX10 0AF.
        Phone: 040-481 4253.

Inglis A S & Son, 43 Davison Villas, Castledawson, Co Londonderry, NI.
        Phone: Castledawson 68548.
Inman H J, 100 Shaldon Crescent, West Park, Plymouth, Devon PL5 3RB.
        Phone: 0752 70822.
Ireland James, 42 Waggon Road, Leven, Fife, Scotland KY8 4RU.

Jackson C A 31 Firle Road, Brighton, Sussex BN2 2YH. Phone: 695287.
Jamieson J T & Son, Violet Bank Cottage, Annan, Dumfriesshire. Phone: Annan 2658.
Jarvis D C, 43 Albert Grove, Ruabon, Wrexham, Clwyd, N Wales LL14 6AF.
Jerome A W, York Cottage, High Street, Lr Wanborough, Nr Swindon, Wiltshire SN4 0AE.
        Phone: 0793 790958.
Johnson & son, No 1 California Terrace, Whitby, North Yorks YO22 4EE.
        Phone: 0947 605530.
Johnston R J, 202 Comber Road, Dundonald, Belfast. Phone: Dundonald 3625.
Jones B, 11 Upper Park Street, Holyhead, Anglesey. Phone: Holyhead 2271.
Jones & Jones, 3 Harcourt Terrace, Brithdir, New Tredegar, Gwent. Phone: 0443 838643.
Jones & Son, 18 Pisgah Street, Kenfig Hill, Nr Bridgend, Glamorgan.
Jones W & G, 31 Eynsham Road, Cassington, Oxford OX8 1DJ. Phone: Oxford 881374.
Keam & Gilhooly, Premier Lofts, 38 Station Road, Whitstable, Kent CT5 1LG. Phone: 274582.
Kehoe Tony, Twin Oakes, Knockmullin, New Ross, Wexford, Eire. Phone: 051-21668.
Kiernan J, 26 Pemberton Road, Bridgehill, Consett, Co Durham. Phone: Consett 503116.
Kelly Paul, 14 Oak Tree Avenue, Llay, Nr Wrexham. Phone: Gresford 5356.
Kennedy R & H, 23 Oaklands Avenue, Irvine, Ayrshire, Scotland KA12 0SE.
King F, 15 Maida Road, Chatham, Kent.

Lawless John, 310 Newton Road, Rushden, Northants. Phone: Rushden 59395.
Lawless Mick, 5 Cam Close, Corby, Northants. Phone: Corby 4660.
Leather Brian, 43 Steeple Road, Mayland, Chelmsford, Essex.
Lee Robert J, 37 Birchgrove, Tirphil, New Tredegar, Gwent, S Wales NP2 6AH.
        Phone: 0443 820838.
Leigh R, Moor View, Bingley Road, Menston, Ilkley, Yorks. Phone;. Menston 2453.
Lewis T, 2 Club Road, Tranch, Pontypool, Gwent NP4 6BH. Phone: 2754.
Lewis Mr & Mrs Wayne, 4 Shroffold Road, Downham, Bromley, Kent. Phone: 01-698 9226.
Lewis Mr & Mrs W, The Freelands, Main Road, Kempsey, Worcester. Phone: 0905-820 924.
Littlewood Keith, The Croft, Shore Hall Lane, Millhouse, Penistone, Sheffield S30 6NG.
        Phone: 0226 763902.

Lowe C G, Effards, St Sampsons, Guernsey, Channel Islands. Phone: Guernsey 4239.
Lowe F, 231 Station Road, Stapenhill, Burton-on-Trent, Staffs. Phone: 48209.
Lycett, Mrs & Sons, 34 Newman Grove, Rugeley, Staffs WS15 1BN. Phone: 0889 583915.
Lyon J, 39 Willow Drive, Skelmersdale, Nr Ormskirk WN8 8PJ.

Maddon & Sons, 15 Crackenedge Lane, Dewsbury, Yorks.
Mair H, 42 Tollerton Drive, Irvine, Scotland. Phone: 76381.
Marshall T, 4 Southward Close, Seaton Sluice, Whitley Bay, Tyne & Wear NE26 4EA.
Martin W, 69 Lytton Road, Leytonstone, London E11.
Matthews Mr & Mrs D, 62 The Crescent, Keresley End, Coventry. Phone: 0203 336689.
McAllister J F, 9 Broughshane Road, Ballymena, Co Antrim, N Ireland BT43 7DX.
McCartney D, Loftwood, Casterton, Kirby, Lonsdale, Cumbria. Phone: 71075.
McCaw W & L, 197 Knock Road, Dervock, Ballymoney, Co Antrim, NI. Phone: 41503.
McCrea Mr & Mrs, Fairfield, 55 Brumer Road, Peterborough PE1 3HT.
McCreight R & E, 14 Straid Road, Ballynure, Ballyclare, Co Antrim, N Ireland.
McKane Mr & Mrs J & Son, Lautoka, 52 Shanard Avenue, Santry, Dublin 9, Eire.
McKay F, 96 Queens Road, Blackburn, Lancs. Phone: 0254 51916.
McLean Jim & Marion, Grahamshill Railway Cotts, Kirkpatrick Fleming, Lockerbie,
    Dumfries DG11 3BQ. Phone: 04618 296.
Micklefield M, 89 Bracken Bank Grove, Keighley, W Yorks BD22 1AU.
Miller, C W & J B, 8 Ffordd Fer, Mynydd Isa, Mold, Clwyd CH7 6XA. Phone: Mold (0352) 58975.
Mills R, 6 Upper Row, Golden Hill, Pembroke, Dyfed, West Wales. Phone: Pembroke 683763.
Minshaw Mr & Mrs & Son, 31 Gatehead, Great Clifton, Workington, Cumbria CA14 9TN.
    Phone: Workington 62469.
Mitchell G E, 39 The Queens Drive, Rickmansworth, Herts. Phone: 092 37 78898.
Montgomery J, Cawood Close, Feltham Avenue, East Molesey, Surrey KT8 9BL.
    Phone: 01-979 6796.
Moore Bros, 44 Boyd Avenue, Kirkcubbin, Co Down. NI. Phone: Kirkcubbin 433.
Morrell W, 1 Front Street, Pelton, Chester-le-Street, Co Durham. Phone: Durham 70043.
Morrison H, 12 Weir Farm, Paddock, Scothern, Lincoln.
Mulholland & Son, 70A Park Road, Downend, Bristol.
Mumford B, 246 High Road, North Weald, Essex.

Neaves N, Cheriton, Eastling, Faversham, Kent. Phone: Eastling 242.
Nelson S, Atlas Lofts, 17 Stokesley Road, Northallerton, N Yorks. Phone: Northallerton 5313.
Nicholson Jim, 5 Mitchell Gardens, South Shields, Tyne & Wear NE34 6EF.
North J A, 49 Foxhills Road, Scunthorpe, S Humberside DN15 8LH. Phone: 863983.

Ord Jas E, 27 Northumberland Place, Barley Mow Estate, Birtley, Tyne & Wear.
Ormond Mr & Mrs & Son, 36 Newlands Avenue, Hartlepool, Cleveland. Phone: 654467.
Osman C A E, 14 Fairgreen, Cockfosters, Barnet, Herts EN4 0QS. Phone: 081-449 8883.
O'Sullivan Gerard, 21 Sundrive Park, Ballinlough, Cork, Eire. Phone: 021 291773.

Page Mr & Mrs J A C W, 'Shangri-La', 18 Blue Cap Road, Stratford-on-Avon,
    Warwickshire CV37 6TG. Phone: 0789 267407.
Parker H G, 22 Pump Lane, Rainham, Kent ME8 7AB. Phone: 0635 373213.
Parkinson & Doherty, 30 Cheetham Meadow, Moss Side, Leyland, Lancs. Phone: 52511.
Parry F P & Grandson, 46 Pleasant View, Wattstown, Rhondda, S Wales.
Partridge N E, 10 Hannah's Field, Ridlington, Oakham, Leics. Phone: Uppingham 823733.
Paterson & Cannon, 30 Kirkoswald Road, Maybole, Ayrshire, Scotland. Phone: 82314.
Patzer A, 27 Desborough Road, Hartford, Huntingdon, Cambs. Phone: Huntingdon 52751.
Pawson F & Son, 27 Legion Street, South Milford, Leeds. Phone: 0977 684264.
Peacock W E & C W, 12 Longley Road, Tooting, London SW17 9LL. Phone: 081-672 4807.
Penfold Mr & Mrs, Jubilee House, Polgooth, St Austell, Cornwall.
Peters & Sons, Hollywood Lofts, 19 Heath Road, Hollywood, Birmingham B47 5LR.
Phillips R, 1 Dan-y-Deri, Woodlands St, Mountain Ash. Phone: 474335.
Pigeonline, John Gore, 35 Kirby Road, Basildon, Essex SS14 1RX. Phone: 0268 583581.
Piper Bros, 51 Geoffrey Avenue, Widley, Nr Portsmouth. Phone: Cosham 7158.
Platts J, 203 Central Road, Hugglescote, Coalville, Nr Leicester.
Pogonowski M A, 10 Sorbus Drive, Crewe, Cheshire. Phone: 583525.
Poole Bros, 18 St John Street, Malmesbury, Wilts. Phone: 066-62 2750/3693.
Porritt J, 63 Park Parade, Westown, Dewsbury, W Yorks WF13 2QL.
Powell R A, 96 Hastings Road, Swadlincote, Burton-on-Trent. Phone: 214084.
Preddy V, 5 Scumbrum, High Littleton, Nr Bristol, Avon.
Price J E, 20 Mallard Way, Moreton, Wirral. Phone 051-677 4852.

Price T C, 8 Booker Lane, High Wycombe, Bucks HP12 3UZ. Phone: 0949 29790.
Proffitt Denis E, 97 Aldershot Road, Guildford, Surrey GU2 6AJ. Phone: 0483 61551.

Quinn J A & Son, 122 Fairlawn Park, Lower Sydenham, London SE26 5SB.
    Phone: 081-659 0880.

Ransome Mr & Mrs V A 'Wiley Flight Lofts', 82 Broadway West, Dormanstown, Redcar,
    Cleveland TS10 5PA. Phone: 0642 475191.
Reeve Bill, Corby Loft, Corby House, East Rainton, Tyne & Wear DH5 9QN.
    Phone: 091-512 0514 or 091-456 6966.
Reynolds J W, Homer, 47 Staffa Drive, Ballymena, Co Antrim, N Ireland BT42 4EH.
    Phone: 0266 44683.
Reynolds Ronald, 8 Middlefield, Weston Turville, Bucks HP22 5RH.
    Phone: Stoke Mandeville 3923.
Rhead Norman, High Trees, Hilderstone, Stone, Staffs.
Richards J, 20 Bourne Ave, Laindon, Basildon, Essex SS15 6DX. Phone: 0268 418015.
Riding Walter, 39 Manor Road, Fleetwood, Lancs. Phone: 77940.
Rimmer R Snr, 87 Pennard Avenue, Huyton, Liverpool. Phone: 051-480 7754.
Roberts Bill, 16 Queens Drive, Larbert, Stirlingshire, Scotland. Phone: 0324 554334.
Robinson G, 159 Lord Street, Grimsby, South Humberside DN31 2JQ.
Robinson John R E, 31 Victoria Avenue, Newtownards, N Ireland BT2 3EB. Phone: 816356.
Rosbotham Ken, 134 Tavanagh Street, Donegall Road, Belfast 12. Phone: 241094.
Ross W & A, 86 Hilton Street, Aberdeen, Scotland. Phone: 44726.
Ruffell Eric, 13 High Road, Chadwell Heath, Romford, Essex.

Sargeant D, Supreme Lofts, Ford Green Mills, Smallthorne, Stoke-on-Trent ST6 1NG.
    Phone: 0728 545484.
Seaton Mr & Mrs & Son, 20 Venables Road, Guisborough, Cleveland TS14 6LG.
    Phone: 0287 635767.
Scott John, 119 Church Lane, Goldington, Bedford MK41 0PW.
Shepherd Mr & Mrs M (NFC Secretaries), 4 Nutcroft, Datchworth, Nr Knebworth,
    Herts SG3 6TG.
Shipley A, The Slack, Crowle, Nr Scunthorpe, Lincs. Phone: Crowle 363.
Sinclair Mr & Mrs Gordon, 21 Corthan Place, Aberdeen, Scotland AB1 5AS. Phone: 875957.
Skewes John H, Delmotte-Jurions, c/o K R Skewes, 64 Dolcoath Road, Camborne,
    Cornwall TR14 8RP. Phone: 0209 714343.
Smilie Bros, 75 Scott Place, Fauldhouse, West Lothian, Scotland. Phone: 229.
Smith D, 11 Hawthorne Avenue, Norton, Doncaster, S Yorks DN6 9HR.
Smith J, 3 Manley Close, Bridgehall Estate, Stockport, Cheshire. Phone: 061-429 6261.
Smith K & P K, Woodlands View Lofts, Pump Lane, Doveridge, Derbyshire DE6 5LX.
    Phone: 08893 66327.
Smith W A, 35 Cleaver Road, Basingstoke, Hants RG22 6SO.
Smyth Bros, 7 Parkmount Avenue, Larne, Co Antrim, N Ireland. Phone: 5884.
Stables Les, Darfield Lofts, 41 Colbourne Street, Swindon, Wilts SN1 2EG.
Stediford J, 21 Clinton Avenue, East Molesey, Surrey KT8 0HS. Phone: 01-979 3676.
Stent Mr & Mrs, ''Corriemore'', Gors Road, Burry Port SA16 0EL. Phone: 05546 2124.
Stephenson W, 21 Christys Lane, Shaftesbury, Dorset. Phone: 0747 2662.
Stevens H J, 98 Humber Road, Coventry, W Midlands CV3 1BA. Phone: Coventry 456280.
Stewart Bros & James, 44 Bond Close, Southwick, Tyne & Wear SR5 1ES.
Stout Mr & Mrs, Duva Lofts, Fir Tree Close, Forest Town, Mansfield, Notts NG19 0JE.
    Phone: 0623 633365.
Sturch W J, 36 Granny Lane, Mirfield, Yorks. Phone: Mirfield 493018.
Sutherton G, Beech Grove, Blaxton, Doncaster DN9 3AF. Phone: 0302 770535.
Sweet S, 27 Blackweir Terrace, Blackweir, Cardiff CF1 3EQ. Phone: 395099.

Telfer R & Son, Cara Columbaria, 16 Neilsland Drive, Motherwell. Phone: 0698 65221.
Thompson, D 57 Whinside Stanley, Co Durham DH9 8AT.
Todd A, 'Todgrove Lofts', 1 Mitchell Stret, Sowerby Bridge, W Yorks. Phone: 0422 835322.
Towers S, Danesbury, Mayfield Flats, Cross-in-Hand, Heathfield, East Sussex.
    Phone: Heathfield 2755.
Townrow H, Chester Lynes, Irchester, Wellingborough, Northants. Phone: 3739.
Tripp John C, Hillyfields Farm, Buckhurst Lane, Wadhurst, East Sussex TN5 6JY.
    Phone: 089-288 2458.

Uttridge J H 26 Laird Avenue, Grays, Essex RM16 2NP. Phone: 0375 383840.
Uzzell A & Son, 45 Henley Drive, Highworth, Swindon.

Venner R & M, 28 Brooks Road, Street, Somerset. Phone: Street 42426.
Vowles E J, Withinfield, 61 St Marys Road, Meare, Nr Glastonbury, Somerset.
  Phone: Meare Heath 208.

Wakerley J H & Son, 60 Woodthorpe Road, Loughborough, Leicester. Phone: 0509 30480.
Ward John, Tyn-y-wern House, Aberkenfig, Mid Glamorgan CF32 9RG.
Warner C, 23 Spencer Street, Gwmanan, Aberdare, Mid Glamorgan CF44 6HN.
Warner G & Son, 22 Ifton Road, Rogiet, Gwent. Phone: Caldicot 0291 422024.
Way Peter, 9 Vine Close, West Drayton, Middlesex. Phone: 47127.
Webster Mr & Mrs, 20 Lloyds Terrace, Cymmer, Nr Port Talbot, West Glamorgan.
  Phone: 850077.
Webster Bros, 20 Lloyds Terrace, Cymmer, Port Talbot, Glamorgan. Phone: 850077.
Wells C, 6 Barnes Close, Beverley, Humberside. Phone: 882488.
West J G, 35 Hill Top Crescent, Harrogate HG1 3BZ. Phone: 503912.
Whitman G F, 303 Southway Drive, Southway, Plymouth PL6 6QN. Phone: 0752 706308.
Wilcox R E, 19 Montague Avenue, Conanby, Conisborough, Doncaster, Yorks.
  Phone: Rotherham 864421.
Williams J A, 22 Beaumont Cresent, Kilvey, Swansea, Wales.
Williamson J & Mason C, Mayfield Nurseries, Ruswarp Lane, Whitby, N Yorks.
  Phone: 0947 602611.
Wilson Michael A, 68 Gloucester Road, Chesterfield, Derbyshire.
Windsor John, 57 Slough Road, Datchet, Berks. Phone: Slough 47607.
Wishart Graeme, 32 Trumper Terrace, Cleator, Cumbria CA23 3DY. Phone: 0946 813378.

FANCY PIGEON ADDRESSES

Feathered World The Editor, 37-39 Fylde Road, Preston PR1 2XQ. Phone: 0772 50246.
National Pigeon Association Sec R Wright, 12 Birdshold Close, Skellingthorpe, Lincoln LN6
  5XF. Phone: 0522 689246. Membership and rings Mary Roscoe, 32 Derwent Drive,
  Prescot, Merseyside L34 2ST. Phone: 051-426 2133.

English Owl Club Sec Bob Snaith, 6 Parkhill Close, Carshalton, Surrey SM5 3QW.
  Phone 081-647 2415.
Fantail Club Hon Sec S E Clayson, 15 Woodside Avenue, Boothville, Northampton.
  Phone: Northampton 499916.
Frillback Club Sec P C J Allen, 26 Manor Park, Histon, Cambridge CB4 4JT.
  Phone: 0223 234481.
Genuine Show & Exhibition Homer Club Sec E Pearson, 538 Manchester Road, Linthwaite,
  Huddersfield, Yorks. Phone: Huddersfield 641243.
German Toy Pigeon Club Sec Claus Liefeith, 526 Huddersfield Road, Wyke, Bradford BD12 8AD.
  Phone: 0274 678910.
High Wycombe Fancy Pigeon Soc Sec R D White, 20 New Road, Bolter End, High Wycombe,
  Bucks HP14 3NA. Phone: 0494 882582.
Ice Pigeon Club Sec J T Roper, 52 West Street, Blaby, Leicester LD8 3GY.
  Phone: Leicester 773865.
Lahore Club Sec Roy Burt, 9 Valley Truckle, Camelford, Cornwall PL32 9RU.
  Phone: 0840 212752.
London and Essex Show Pigeon Soc Sec A Forder, 195 Woodward Road, Dagenham, Essex.
  Phone: 081-595 4174.
Long Faced Clean Legged Tumbler Club Sec John Surridge, 298 St Richards Road, Deal, Kent
  CT14 9LG. Phone: 0304 364935.
Magpie Club Sec J Kitchen, Heathcote Grange, Hartington, Nr Buxton, Derbyshire SK17 0AY.
  Phone: 029 884210.
Midland Columbarian Soc Hon Sec Sidney E Clayson, 15 Woodside Avenue, Boothville,
  Northampton NN3 1JL. Phone: 0604 499916.
Muffed Tumbler Club Sec D & M Conyers, 27 Benningholme Lane, Skirlaugh, North Humberside
  HU11 5EA. Phone: 0964 562600.
National Federation of West Tumblers and Exhibition Flying Tipplers Secs Mr & Mrs R
  Greenwood, 140 Ashover Road, Old Tupton, Chesterfield. Phone: 0246 863415.
National Modena Club Sec Mrs C Wright, 12 Birdsholt Close, Skellingthorpe, Lincoln LN6 5XF.
  Phone: 0522 689246.

Listed below are some of our regular writers for The Racing Pigeon weekly or The RP Pictorial who can be contacted.

FEATURES

CLUB & ADMIN QUERIES, Clubman c/o Unit 13, 21 Wren Street, London WC1X 0HF.

MONTH BY MONTH. F W S Hall, 73 London Road, Enfield, Middlesex EN2 6ES. Phone: 081-363 2125.

OFF PRESCRIPTION, Dr Tim Lovel, Holywell Hall, Brancepeth, Co Durham.

ANOTHER VIEWPOINT, Jack Adams, 1 Willow Way, Redditch, Worcestershire B97 6PH. Phone: 0527 68738.

THE SHOW SCENE, Doug McClary, 59 Argyll Road, Pennsylvania, Exeter, Devon EX4 4RX. Phone: 0392 73947.

YOUNG FANCIERS' FORUM, Jonathan Mee, 178A Kilburn Road, Belper, Derbyshire DE5 1SB. Phone: 0773 824618.

FOOD FOR THOUGHT, Dave Allen, Rose Cottage, West Markham, Newark, Nottinghamshire NG22 0PN. Phone: 0777 872096.

PROF BILL MULLIGAN, 1 Daisy Green, Groton, Near Boxford, Colchester CO6 5EN.

D V BELDING, 5 Brook Close, Ruislip, Middlesex HA4 8AF.

STEVEN VAN BREEMEN, Leeghwaterstraat 68, 1221 bh Hilversum, Holland. Phone: 010-31-35-830 285.

DOUG WENT, 40 Cowper Gardens, Southgate, London N14 4NR. Phone: 081-449 9466.

SCOTLAND

BOB KENNEDY (Culzean), 23 Oaklands Avenue, Irvine, Scotland KA12 0SE. Phone: 0294 75793.

JOHN T IRONSIDE (Ironside), 58 Broomhouse Bank, Edinburgh, Scotland. Phone: Edinburgh 6574 525.

IAN GRAY (Smokie), Woodlands by Forfar, Fife, Scotland. Phone: 0307 65849.

DEREK STREET, 1 Inkerman Terrace, Rothesay, Isle of Bute PA20 0JE. Phone: 0700 502370

J RENWICK, 113 Fergusson Road, Broxburn, West Lothian. Phone: 0506 852900.

W P WORTLEY, 37 Hospital Road, Annan, Dumfriesshire DG12 5JF.

DONNY BROOK, 213 Wishaw Road, Waterloo, Wishaw, Lanarkshire ML2 8ES.

IRELAND

RONNIE JOHNSTON (Irish Rover), 202 Comber Road, Dundonald, Belfast, Northern Ireland BT16 0BS. Phone: 0232 483625.

AIDEN WHELAN (Silver Spurs), Mountain Vista, Moyne Upper, Enniscorthy, County Wexford, Eire. Phone: 054 35354.

SEAMUS LEHANE (Eblana), 69 Lorcan Crescent, Dublin 9, Eire. Phone: 421391.

CLIFFORD BROWNE, 29 Westburn Crescent, Bangor, Co Down BT20 3BN. Phone: 0247 465 273.

J W REYNOLDS, 47 Staffa Drive, Ballikeet, Ballymena, Antrim BT42 4EH.

WALES

JOHN DAVID (Dikke Blauwe Duiven), 1 Llest Cottages, Llantwit Fardre, Near Pontypool, Mid Glamorgan. Phone: 0443 223213.

J G THOMAS (Stag Nights), Whitefield, High Street, Bagilt, Clwyd CH6 6AP. Phone: 03526 734709.

GARETH WATKINS, 44 Eleanor Street, Tonypandy, Rhondda, Mid Glamorgan. Phone: 0443 435684.

W J (BILL) OWSTON, The Lodge, 117 Priory Road, Milford Haven, Dyfed.

MOELWYN DAVIES (Colomen), 1 Hillcrest Drive, Glynfach-Porth, Rhondda, Mid Glamorgan.

JOHN DIXON (Dixey), 73 Brgnyfryd Terrace, Ferndale, Mid Glamorgan CF43 4HT. Phone: 0443 730492.

NORTH EAST

DAVE BARKEL (In The 4 Ground), 35 Rosslyn Avenue, Ryhope, Sunderland SR2 0SB. Phone: 091-521 0810.

JACKIE TRAYNOR, 2 George Street, Craghead, Stanley, Co Durham. Phone: 0207 235042.

JIMMY GRIGG, 5 Front Street, Davy Lamp, Kelloe, Co Durham DH6 4PQ. Phone: 091-377 1364.

TERRY CALLAN, 35 Skelton Road, Brotton, Saltburn-by-Sea, Cleveland. Phone: 0287 77489.

ALF ROTHWELL, 58 Ennerdale Road, Walker Dene, Newcastle upon Tyne. Phone: 091-262 5440.

V A RANSOME (Wiley Flight), 82 Broadway West, Dormanstown, Redcar, Cleveland TS10 5PA. Phone: 0642 475191.

JACK CURTIS, 70 Ryhope Road, Sunderland, Tyne and Wear. Phone: 091-565 9452.

BOB PINKERTON (Pinky), 160 Denton View, Wintaton, Tyne and Wear NE21 4DY. Phone: 091-414 4537.

EASTERN ENGLAND

MELVYN HORN (Henpecked), 34 Windsor Road, Reydon, Southwold, Suffolk IP18 6PQ. Phone: 0502 724473.

JACK BATEMAN (Tithe Rambler), 2 Tithe Road, Chatteris, Cambridgeshire PE16 6SL.

NORTH WEST

J KIMMANCE (Kim), 52 Brunel Drive, Litherland, Liverpool L21 9LP. Phone: 051-928 5946.

LES PARKINSON (Parky), 74 Long Lane, Middlewich, Cheshire CW10 0EN. Phone: 060-684 6183.

BRIAN NEWSON (Ribble Rambler), 20 Gorsey Lane, Banks, Southport, Merseyside.

JOHN PILLING (Pilling's Patch), 26 Fern Close, Skelmersdale, Lancashire WN8 8DN.

A P WRIGHT (Wyre Light), 311 Fleetwood Road, Fleetwood, Lancashire FY7 8AT. Phone: Fleetwood (2611) 03917.

T McGARVEY (Kirkby Mac), 11 Bainton Close, Southdown, Kirkby, Lancashire L32 7PD.

PETER J BRAY (Lancashire Hotpot), 12 Farmers Row, Blackburn, Lancs BB2 4NN. Phone: 0254 55489.

YORKSHIRE

BRIAN COOPER (Blue Cock), 15 Barnburgh Lane, Goldthorpe, Rotherham, South Yorks. Phone: 0709 893492.

HARRY JACKSON (Darkie), Laurel House, Main Street, Hemingborough, Selby, North Yorks. Phone: 0757 038328.

DON GWILLIAM (Sadler), 2 The Bungalows, Ledston Luck, Kippax, nr Leeds LS25 7ET.

R BAXTER-DORE, 40 Hainworth Lane, Ingrow, Keighley, West Yorks. Phone: 0535 681396.

TONY OATES, Two Trees, Knott Lane, Rawdon, Leeds LS19 6JL. Phone 0532 503139.

D SWALES (The Brooker), 71 Eldon Road, Eastwood, Rotherham, South Yorks.

MIDLANDS

R D BEARDS, 11 Chestnut Avenue, Tipton, West Midlands. Phone: 021-520 4972.

MARK PALMER, 4 Conway Grove, Hamstead, Birmingham B43 5HD. Phone: 021-358 1267.

JOHN GWYNNE (Hawkstone), Laburnam Cottage, Bear Croft, Hinstock, Market Drayton, Shropshire TF9 2SZ. Phone: 095 279341.

RON JACKLIN (Leen Sider), 31 Bunnerman Road, Bulwell, Nottingham NG6 9GA. Phone: 0602 752093

MARK EVANS, 14 Coronation Road, Ocker Hill, Tipton, West Midlands DY4 0YA.

HOME COUNTIES

BOB ARCHER (Barwick Green), Trevisky, Cornwall TR16 6AT.

RICHARD BOYLIN (Boylin Points), Blackthorn Cottage, Tyringham, Newport Pagnell, Bucks MK16 9ES. Phone: 0908 612517.

LONDON & SOUTH EAST

KEVIN FOSTER, 152 Old Road East, Gravesend, Kent. Phone: 0474 358599.

BRIAN WOODHOUSE, 1 Overton Drive, Wanstead, London E11. Phone: 081-530 4758.

WARREN FOSTER, 4 Valley View, Greenhithe, Kent. Phone: 0322 845541.

KEITH MOTT, 38 Foxwarren, Claygate, Esher, Surrey. Phone: 0372 67185.

GEORGE BURGESS (Pigeon George), Rosemead, 8 The Drive, Wraysbury, nr Staines, Middlesex TW19 5CS. Phone: 0784 813662.

DOUG CRANSTOUN, 12 Stodart Road, Penge, London SE20.

JEAN ANDREWS (M11), Willow Tree Cottage, Green Tye, Near Much Hadham, Herts SG10 6JH. Phone: 0279 842831.

SOUTH & SOUTH WEST

JANET CRAWFORD, 107 Rutten Lane, Yarnton, Oxford. Phone: 08675 6668.

ALAN CHAFFEY (Bristol Bulldog), 214 Bedminster Road, Bedminster, Bristol. Phone: 0272 668111.

JACK BOLITHO (Trelawney), 19 Oates Road, Helston, Cornwall. Phone: 0326 573835.

RICHARD GREEN, Middle Cottage, Sutton End, Crockerton, nr Warminster, Wilts BA12 8BQ. Phone: 0985 217574.

CATHERINE EMERY, 14 Vale View, Radstock, Avon BA3 3QD. Phone: 0761 36028.

HARRY BENNEY (Plymothian), 50 Cotehole Avenue, Keyham, Plymouth, S Devon PL2 1LU.

PIGEONLINE

JOHN GORE, 35 Kirby Road, Basildon, Essex SS14 1RX. Phone: 0268 583581.

CONTENTS

REFERENCE SECTIONS

INDEX TO ADVERTISERS

Racing Pigeon Books and Sundries Price List . . . page 416